The Birdman of Koshkonong

THE BIRDMAN
OF KOSHKONONG

The Life of Naturalist Thure Kumlien

MARTHA BERGLAND

WISCONSIN HISTORICAL SOCIETY PRESS

Published by the Wisconsin Historical Society Press
Publishers since 1855

The Wisconsin Historical Society helps people connect to the past by collecting, preserving, and sharing stories. Founded in 1846, the Society is one of the nation's finest historical institutions.
Join the Wisconsin Historical Society: wisconsinhistory.org/membership

Front cover images: Painting of a white wagtail by Thure Kumlien, courtesy of Gregg Kumlien; greater prairie-chicken feather, courtesy of mycteria/Shutterstock.com; other specimens, clockwise from top left: red-winged blackbird eggs, least flycatcher nest with eggs, and black tern egg, courtesy of University of Wisconsin Zoological Museum. All nest and egg specimens on the front and back covers were collected by Thure Kumlien.

Spine image: Red-winged blackbird egg, courtesy of University of Wisconsin Zoological Museum.

Back cover images: Sandhill crane feather and hummingbird nest with eggs, courtesy of University of Wisconsin Zoological Museum; *Aster borealis* (top) and *Solidago* L. (bottom) collected by Thure Kumlien, courtesy of the Wisconsin State Herbarium.

Dedication, page v: This landscape was painted by Gregg Kumlien, great-great-grandson of Thure Kumlien. In his work, Gregg says he tries to "rekindle the magic and remembered beauty of the Wisconsin of my youth." Courtesy of Gregg Kumlien.

Printed in the United States of America
Cover designed by Steve Biel
Typesetting by Integrated Composition Systems
25   24   23   22   21   1   2   3   4   5

Library of Congress Cataloging-in-Publication Data
Names: Bergland, Martha, 1945– author.
Title: The birdman of Koshkonong : the life of naturalist Thure Kumlien /
    Martha Bergland.
Description: Madison : Wisconsin Historical Society Press, [2020] |
    Includes bibliographical references and index.
Identifiers: LCCN 2020034666 (print) | LCCN 2020034667 (ebook) |
    ISBN 9780870209529 (paperback) | ISBN 9780870209536 (ebook)
Subjects: LCSH: Kumlien, Thure Ludwig Theodor, 1819–1888. | Naturalists—
    United States—Biography.
Classification: LCC QH31.K86 B47 2020 (print) | LCC QH31.K86 (ebook) |
    DDC 508.092 [B]—dc23
LC record available at https://lccn.loc.gov/2020034666
LC ebook record available at https://lccn.loc.gov/2020034667

♾The paper used in this publication meets the minimum requirements of the American National Standard for Information Sciences—Permanence of Paper for Printed Library Materials, ANSI Z39.48-1992.

To Jim and to Brita

Published with the support of
MILWAUKEE AUDUBON SOCIETY
*protecting and restoring Wisconsin's natural heritage and ecology
through active leadership, education, advocacy, and stewardship since 1897.
Milwaukee Audubon Society and the Wisconsin Society for Ornithology
owe their early strength to the pioneering of frontier scientist and birdman,
Thure Kumlien.*

In his early days in Wisconsin, Thure Kumlien discovered a bed of *Linnaea borealis*, or twinflower, in the tamarack swamp bordering Lake Koshkonong. NEW YORK PUBLIC LIBRARY VIA HATHITRUST AND GOOGLE DIGITIZED

CONTENTS

A Note on Terminology

Many bird names—both common and scientific—have changed since Thure Kumlien's day. Because untangling the outdated names from those that have not changed would present an unnecessary challenge to most readers, I have used the modern common names for birds throughout the book.

In direct quotations from Kumlien and others, and in instances where I paraphrase original quotations, I have included the modern common names in brackets after the names used in the original text (eg. "quail [northern bobwhite]"). Additional notes regarding common and scientific bird names also appear in certain endnotes.

Thanks are due to R. Tod Highsmith—former officer of the Wisconsin Society for Ornithology and former editor of its journal, *The Passenger Pigeon*, as well as an enthusiast of the history of ornithology—who assisted in identifying archaic usage throughout the book and providing modern translations.

Translating Swedish to English posed some challenges in the creation of this book, as the Swedish alphabet has more characters (three more vowels: å, ä, and ö) than the English alphabet. I have kept Swedish place names, as well as Swedish first and last names, in Swedish. However, I have standardized both Swedish and American proper names that appear in various spellings throughout primary source documents. Some endnote citations have not been translated from Swedish into English, because no English translation of those sources exists.

Thanks are due to Lena Peterson Engseth—archival and genealogical researcher, librarian, and poet—who provided all translations and transcriptions of Swedish text for this book, as well as this pronunciation of Thure Kumlien's name for English speakers: TOO-reh koom-LEEN.

PROLOGUE

In 1888, Edward Lee Greene remembered a day thirty years earlier, a summer day in about 1858, when as a half-grown boy he walked along a road through rolling farmland near Lake Koshkonong in Wisconsin. After a mile or so, he stepped off the straight road onto a path that wound through old oak woods. He was on his way from his family's farm to the home of his friend, Mr. Kumlien.[1]

Edward was passionate about the study of plants, and Mr. Kumlien was especially knowledgeable in that field. He knew the common and the rare, the beautiful and the insignificant, the useful and the poisonous. He had studied botany at the same university in Sweden where the great Carl Linnaeus had taught. Then he had come to America in 1843. Edward believed Mr. Kumlien knew more about birds, bird eggs, and bird nests than anyone in Wisconsin.

Farmers who were not distracted by the study of the natural world had prospered on this rich land. They raised wheat and corn in square fields and lived in frame houses next to the roads. But Mr. Kumlien still lived like a settler in a log cabin in the woods on an Indian trail.

On a ridge at the edge of the woods, Edward stopped a moment to look down at the creek wandering through the wet meadow to the shore of the lake. Above the blue line of the lake, he could see low hills to the southeast. As he continued along the path to the cabin and opened the first gate, he thought again what a perfect home this was for a poet and naturalist—pristine, secluded, and quiet.

In the early spring of 1858, when the ice had melted, Edward and Mr. Kumlien had visited the three little lakes nestled in the wooded hills to the west. They had climbed the bluffs to see the earliest spring flowerings in the thin soils and rocks—the purple of the pasqueflower, the prairie buttercup, the almost insignificant *Draba caroliniana*, and the *Arabis lyrata* draping small white flowers along the bleak summit. Now, in the summer, the reedy margins of those lakes would be blooming with *Pontederia cordata*, which the farmers call pickerelweed, and the floating *Brasenia*, which blooms deep red for just two days. There would be every kind of water lily—most abundantly the fragrant white *Castalia tuberosa* and the very beautiful yellow-flowered *Nelumbo lutea*, sometimes called the water-chinquapin because, like the chinquapin oak, it bears edible seeds. Each time he went off on a walk with Mr. Kumlien, Edward's knowledge of the world became richer in beauty, detail, and delight.

But on this summer day, Edward and Mr. Kumlien set off up the road, walking between fields of wheat that had been unplowed prairie just a dozen years earlier. They headed north to a little tamarack swamp, very rare this far south, in a low place surrounded by heavy forest. In the tamaracks, they would look for the ericaceous, acid-loving plants and orchids not found in prairies and timbered upland.

In a field of green wheat near the crest of a hill, Mr. Kumlien pointed out an almost invisible pair of sandhill cranes. *Trana*, he called them in Swedish, and he sang out what sounded like a verse of Swedish poetry. Edward, who spoke Norwegian, thought Mr. Kumlien was telling the cranes that they were safe, for today the two friends hunted plants, not birds.

Mr. Kumlien, who had lived in this country only fifteen years, knew so well the varied flora of southern Wisconsin that the scientific name for every tree and shrub, flower, grass and sedge and moss, lichen and mushroom seemed to be at his tongue's end. And the plants were not even what he knew best. Though he was a man of short stature with rather stooping shoulders, he had a surprisingly muscular frame. With his light hair and blue eyes, a stranger who passed him on the road might mistake him for an ordinary Scandinavian farmer. But this farmer could write Ciceronian Latin and address a foreigner in many of the languages spoken between Spain and Sweden.

As they entered the swamp, the tamarack, which had never been cut, formed an almost impenetrable forest. Edward tried to stay close in order to see and hear what Mr. Kumlien had to say about the flora. In this little swamp, Mr. Kumlien identified plants never found in the upland, on the big lake, or even at the little lakes. He named the gray lichens that draped the older branches of the tamaracks while they stood on the uneven mat of lichens and mosses. He spotted cranberries and blueberries with berries forming. He pointed out patches of leatherleaf shrub with its little white bells and the wintergreen at their feet.

Gaulteria, Mr. Kumlien told Edward, was named after a Mr. Gault whom the Swedish collector Pehr Kalm met in the 1740s on his North American sojourn. Pleased with his best student Pehr Kalm, Linnaeus agreed to honor Kalm's friend with the name. Edward could see that these ericaceous shrubs were a delight to this man whose early home had been in northern Europe.

At the edge of a sunny opening in the swamp, Mr. Kumlien stopped and waited for Edward to stand beside him. They quietly took in the bright, almost birdlike blooms nodding in the sun. There, like a white bird carrying a pink sack, was the showy lady's slipper on its stout and hairy stalk. There, on its slender stem, was the graceful rose pogonia with its bearded and fringed labellum and a scent like raspberries. And the *Arethusa bulbosa*, or dragon's mouth orchid, Mr. Kumlien called the love- liest of all North American wildflowers. Known for its exquisite form and its rich scent, this *Arethusa* presented one graceful pink flower on each slender, leafless stem, the white labellum bearded yellow and spotted with magenta.

In this little bit of botanical paradise, Mr. Kumlien recalled, he once discovered a bed of what was dearest of all to the hearts of Swedish bota- nists, *Linnaea borealis*, the favorite of Linnaeus, who put this twinflower on his coat of arms. It was a low, trailing plant with each set of small round leaves strung along a delicate stem, and from each leaf site came a dark red, wiry stem holding up the twin pink bells. Mr. Kumlien had made this discovery in his early days in Wisconsin, and he never found the *Linnaea* again. Though they searched, Edward and Mr. Kumlien did not find the twinflower that day either. But many years later, when he was a man, Edward had the satisfaction of carrying a sprig of *Linnaea* from what must have been the original and long-lost spot to his friend Mr. Kumlien. Edward never found it again after that, though he searched for it time after time.

The last letter that Edward Lee Greene received from his friend Thure Kumlien many years later was tinged with melancholy as he related how this long-cherished tamarack swamp near his home had been destroyed.

Its trees had been cut down, as had the ericaceous undershrubs and the delightful orchids. A farmer had drained it and plowed it and planted it with market-garden vegetables.

This was one sadness among the many joys, sorrows, and wonders of a life fully lived, a life that began in the richness of Mr. Kumlien's beautiful Sweden.

INTRODUCTION

Few people have heard of Thure Kumlien, a Swedish American settler who studied birds and plants in the unspoiled forests, swamps, and prairies near Lake Koshkonong in Wisconsin. He was not even well known in Jefferson County where he lived from 1843 until 1888. But anonymity did not worry Thure Kumlien. Fame is not what he sought so avidly.

Kumlien was born in Sweden in 1819. His well-educated upper-class parents encouraged his early interest in natural history and provided him with a fine education. At Uppsala University, he was befriended by the scholar and botanist Elias Fries, who then occupied the chair once held by Carl Linnaeus. Kumlien might have followed in the footsteps of Fries at the university, but he met and fell in love with Christine Wallberg, a "serving maid." The couple immigrated to North America with Christine's elder sister, Sophia, arriving in Milwaukee in August 1843. Married that September in Milwaukee, Thure and Christine settled on the shore of Lake Koshkonong where, after seven years spent enduring the hardships of the settler's life, their luck changed, their lives eased, and they began raising a family. Kumlien farmed, but his eye was on the birds. He continued the collecting and taxidermy he had begun as a child, eventually augmenting his meager income by supplying specimens of birds, eggs, and nests to European and American museums and collectors, including Louis Agassiz, who said no one knew more about bird nests than Thure Kumlien. Kumlien taught languages and science for a number of years at nearby Albion Academy, an important frontier school. When the school closed, Kumlien was hired to gather specimens for the young University of Wisconsin's natural history collection. He later became the third employee and first curator of the Milwaukee Public Museum. He was working for the public museum when he died in 1888.

Why have most people not heard of Thure Kumlien? He had many attributes that, in combination with his accomplishments, might have made him "famous." He was born into wealth in an old, well-educated, well-read family. He had the artistic ability, even as a child, to draw and

A white wagtail (*Motacilla alba*) painted in Sweden by Thure Kumlien. COURTESY OF
GREGG KUMLIEN

paint the birds he knew in Sweden. He spoke many languages and often
played the flute for family and friends. As a young man, he was confident
enough of his own mind and judgment that he immediately knew he had
met the love of his life when he met Christine Wallberg—and he was right.
When it was clear to him that he and Christine could not make a place
together in Sweden, they quickly decided that they should immigrate to
North America. In 1843, this immigrant road was not a well-traveled one.
Only a few Swedes had made the same momentous choice. So Kumlien
was also a bold man. Why isn't this bold, scholarly, talented man more well
known? Because other attributes that often accompany these traits were
simply not important to him.

   A gull, an aster, and an anemone are named after Kumlien, yet there
are no schools or parks named after him, as there are for naturalist Increase

Allen Lapham (1811–1875). Lapham was a quiet and unassuming man, like Kumlien, but Lapham was a public figure as well as a naturalist. It was important to Lapham that he be credited for his work, and much of his work was in the public arena as a citizen. He wrote to record and to achieve ends for the common good. Kumlien, by contrast, was a very private citizen whose work was almost always behind the scenes. His bird skins and mounted specimens could be seen in museum displays, though they more often existed behind the scenes at museums in scientific collections. His neighbors knew him as a man obsessed with birds, but many of them didn't know he was corresponding with naturalists all over the world and sending bird specimens to Boston, New York, and Washington, DC, and to Stockholm, Munich, and Amsterdam.

We know a great deal about later naturalists John Muir (1838–1914) and Aldo Leopold (1887–1948). They, like Lapham, had much to say about the despoiling of the natural environment—the same changes Kumlien witnessed firsthand. But Lapham, Muir, and Leopold were all writers; it was important to them to be known as writers. We know of them because their many published writings live on after them. Kumlien wrote many letters and a diary, as well as lists and descriptions of birds, but most of his writing was not for the public. He wrote several articles, but he declined to put his name on at least one of them. His actions are difficult to understand in today's world, where it's good to get your name and face out there, where if you're not going to be rich, you should at least try to be famous.

Thure Kumlien is difficult to know for other reasons as well, due partly to the normal ravages of time and partly to the difficulty his son must have had in dealing with his father's voluminous, unsorted, unlabeled papers after Kumlien's death. Most were written in Swedish, which the family by then did not speak. Kumlien's granddaughter Angie Kumlien Main, who was about five when he died, nevertheless remembered him and did more than anyone else to keep his memory alive by writing several memoirs of his life using the papers that remained after his death. But when she was a child, those papers were in disarray, a biographer's nightmare. She wrote: "I loved to go into the big room, or loft . . . where on the floor was a huge pile of letters. As I look back and remember how I, as a small girl, used to climb over them, it seems as though there must have been a wagon-box full."[1] Unfortunately, no wagon box of Kumlien's papers exists today. Only a small portion of his papers—which Angie described as "Grandfather's most prized letters from relatives, friends, teachers, scientific correspondents, together with his personal papers"—survived along with a journal he kept in Swedish between 1844 and 1850. These were the papers Kumlien "kept in an old oak chest with hand-wrought iron handles and trimmings, which he brought with him from Sweden."[2] The chest still exists at the Hoard Museum in Fort Atkinson, Wisconsin. Kumlien's papers are primarily at the Wisconsin Historical Society Archives in Madison: three boxes of papers rather than a wagon-box full.

This dearth of Kumlien's written work is not simply the fault of descendants who didn't know how to care for the documents. Kumlien probably had not labeled or sorted most of his papers. He had indicated that some of them were important by putting them in the oak chest, but the others outside the chest, the ones whose stamps had been torn off by grandchildren, were likely discarded. Main remembered sitting on the stack of papers as a girl and reading her grandfather's letters while bees and hornets buzzed around her.

Though Kumlien came to know more than almost anyone at the time about the birds of Koshkonong, he didn't write about them beyond keeping field notes and writing letters. He talked about birds and plants to his children, to the gifted neighbor boy Edward Lee Greene, and to a few interested neighbors. He had a great deal to say to people he knew. Neighbors, friends, and acquaintances were so drawn to him—his recitation of poems, his flute playing, and his cabin full of bird specimens—that he sometimes complained about too many visitors, how their presence hindered his preservation work and his work on the farm. He was called shy, but people often mistake quietness for shyness. Kumlien was a slight man who did not put himself forward aggressively in his voice or in his person. Yet he was not lacking in what we now call self-confidence. He was not a man easily known—not by the people who knew him during his life and not by us now.

The facts of Kumlien's early life can be put in a teacup. We have no letters he wrote when he was at Uppsala, no early impressions of the place he grew up. In order to tell the story of his life in Sweden, we have to look at the worlds he lived in during the 1820s, 1830s, and 1840s—his family history, the house he lived in, the land he wandered, his early schools, his university. It is from his later writings, from what people have said about him, and from the influence he had on others that we get a sense of his personality and his temperament.

Thure Kumlien was not well known during his lifetime nor after his death. His work was quiet work. So, why write a biography of this very private citizen? Because his quiet work was hugely important work in the fields of ornithology and botany. Because there is something romantic, mysterious, and magnetic about him. Because, though he lived the hard life of a settler in the Wisconsin Territory, he was not like other settlers. He was a poetic man of the Koshkonong forests, swamps, and lake.

I am not the only writer who has been drawn to Kumlien. I first saw his name as the title of a poem by Wisconsin's great poet Lorine Niedecker (1901–1969). Niedecker was a lifelong resident of Blackhawk Island along the Rock River at the head of Lake Koshkonong. She, too, lived a quiet life, making Koshkonong a subject of her work. She was attracted to the quietness of Kumlien's life, its simplicity compared to the passion he applied to his studies, and its poverty compared to the wealth of, in her words, "the bigwigs" from Boston who bought his birds. Niedecker's untitled poem, which includes words from Kumlien's journal, hints at the ambivalence he might have felt about hunting birds:

Shut up in woods
he made knives and forks
fumbled English gently:
Now is March gone
and I have much undone

It would be good
to hear the birds
along this shore intently

without song of guns.[3]

Unfortunately, Kumlien, along with most other early ornithologists, had to kill the birds he wanted to study. In the nineteenth century, there was no other way to bring a bird close enough to identify it, measure it, or paint it. Without cameras or binoculars, the only way Alexander Wilson (1766–1813) and John James Audubon (1785–1851) could study birds up close was by shooting them or trapping them in nets. All of those magnificent nineteenth-century birdmen killed a lot of magnificent birds. Niedecker must have seen the beauty in this tension and assumed that it existed within the gentle Kumlien.

A prolific writer of midwestern lakes, rivers, and lore, Walter Havighurst (1901–1994) wrote a fine novel based on the life of Thure Kumlien. In this 1940 book, *The Winds of Spring*, a young scholar and naturalist, Jan Carl Sorensen, meets and falls in love with a girl beneath his station, and the two immigrate from Sweden to Lake Koshkonong. Sorensen, like Kumlien, must farm to make a living, but his heart is in the study of nature. On the rainy night when he and his bride arrive at their land, Sorensen slips out of the wagon where they are sleeping.

> That night he walked miles over a silent country, through deep woods where the brush rained on him, past meadows wet and gleaming under the moon, across marshes where he sank to his knees in carex swamp and sponge moss. He circled sleeping farms, tight squares of darkness in a silvered field. He climbed rail fences where the cattle stood like moonlit boulders, their big ears cupped to catch his passing. He passed without a sound through gleaming thickets where green eyes like twin glowworms watched him from the dark. He walked like a man of iron energy. . . . He began to see that it was infinitely difficult in the place that he had chosen. There must be a world outside the daily boundaries. He must have access to knowledge, and to mystery.[4]

It's not difficult to imagine Sorensen as a perfect stand-in for Kumlien, wandering his land and hungering for a deeper understanding of its inhabitants.

The Wisconsin children's writer Sterling North, whose father's property adjoined the Kumlien homestead, recounted this story in his memoir *Rascal*:

"Kumlien could start the whippoorwills any night by playing his
flute," my father said. "Far across the fields, we heard them, the
old man with his flute, his son playing the violin, and hundreds of
whippoorwills calling—that's music to remember." It made me sad
that I could not have known Kumlien, and walked the woods with
him, learning every bird and flower and insect. [5]

The writers who are drawn to Thure Kumlien are drawn to his aura of
sadness, to his sense of beauty, and to his drive for knowledge—all con-
trasted with his poverty and the unrelenting and physically demanding
work of the settler that encompassed his first seven years on the land.
Kumlien was not like most of the settlers around him. The mystery of his
character pulls us toward him.

Kumlien was well loved in his life—by his wife of many years, Christine, by his four children, and by many friends. The dedication and fore-
sight of three of those who loved him make it possible to tell Kumlien's
story—Ludwig Kumlien, Edward Lee Greene, and Angie Kumlien Main.
Thure inspired, taught, and encouraged them—son, neighbor, granddaughter—into successful lives as naturalists. Their stories are part of his
story. Each one of the three wrote to him and about him, and they saved
his letters and many other documents. In many ways, they are the ones
responsible for passing him on to us.

What Kumlien required of life was *beauty* and *discovery* and *knowing.*
If, on some April evening, he heard silvery notes off in an oak opening, he
not only had to identify the brown little speckle-breasted bird as the migrating hermit thrush but he also had to bring it close and hold it in his
hand. Being known in Wisconsin the way he was in Europe and the East
was not important to Kumlien, but Wisconsin will be richer for knowing
more about this subtle citizen.

This is the story of a "charming, scholarly fine-grained man," [6] a bold
young Swedish immigrant, a poetic and driven scientist, a narrow-
shouldered suitor, a settler who had never swung an axe, an excellent yet
reluctant teacher, and a museum curator who died a tragic death. Though
looking for Thure Kumlien is in some ways like looking for the hermit
thrush, his story can be brought close through his Swedish work journal,

through drafts of letters he sent, through letters he received and saved, and even through his lists and scraps of notes. But his story is most vivid to us when we see through his eyes the lands he wandered and studied and hunted—the remnants of forest, savanna, and marsh on the shores of Lake Hornborga and Lake Koshkonong.

1

A Swedish Son and Schoolboy

1819–1839

In the heart of southern Sweden is a large shallow lake where, every spring, thousands of Eurasian cranes stop on their migration north to rest and to dance. Mated pairs of man-high white birds leap and bow with wings wide, stretching their black and bright-white necks, showing off messy, plumy tails and bright red caps. The raucous din of the great flocks' harsh cries and their muttering and purring on the ground are as impressive as the bird's stature and wingspan. Each year, thousands of birders and tourists travel to Lake Hornborga to see and hear the return of the cranes from the south. Not only cranes make this stop on their migration—so do mute swans, whooper swans, graylag geese, and many varieties of ducks, waders, and warblers.

Since the retreat of the last glacier about ten thousand years ago, this ten-square-mile lake in a valley of Silurian limestone has fed and sheltered an astounding abundance of birds, fish, and mammals. And for at least a thousand years, people have lived on the shores of Lake Hornborga where they farm, fish, and hunt. Since the 1800s, Swedish naturalists have been drawn to the lush, noisy life of the birds of Hornborga.

The storied lake, a bird-collecting schoolboy's dream, was about four miles from Thure Kumlien's childhood home on the Flian River, which flows out of Lake Hornborga. We don't have Thure's impression of the lake, but another schoolboy naturalist, sixteen-year-old Gustaf Kolthoff,[1] made his first visit to the lake in about 1865, noting: "As a child I had listened

The water lilies, reeds, and grasses near the shore of Lake Hornborga, pictured here ca. 1900, attract thousands of migrating and breeding water birds to south central Sweden. PHOTO BY KARL FREDRIK ANDERSSON, VÄSTERGÖTLANDS MUSEUM

with close attention to people's stories about the bird life at Lake Hornborga." Kolthoff told of hiking from his home to the lake over hills and plateaus, through fields and wetlands. Kumlien, too, must have come out of the forest to see Lake Hornborga spread before him, wide and welcoming. "I heard strange bird song and saw in the distance unknown birds circling," Kolthoff recalled.[2] In order to see with his own eyes "all these unknown creatures" whose sounds he heard in the "secretive sedges," he borrowed a boat. It was, he said, a lovely evening, the sort of evening when a lover of nature would rather "go to the woods than go to bed."[3] He heard as much as he saw in the dusk: "Once in a while a mallard flew up from the reeds, screaming, and a multicolored male northern shoveler flew swinging by, as if he wanted to show how beautifully his blue wings were gleaming" in the slanted light. "Common teals, fearless, paddled close . . . in the clear water," he wrote, "and a pair of tufted ducks flew by murmuring in a wide turn over the reeds and came back to the place they had chosen for their nest; a grassy turf beside a clear water puddle."[4] The next morning, the young naturalist shot some of the birds for his collection, wrapping them in paper and putting them away in the tin vasculum he used also for his botanical collecting.

As a young boy, Thure Kumlien, too, studied and collected birds—in the woods and fields and marshes, on Lake Hornborga, and on its shores near his home in Härlunda, Västergötland, not far from Skara in south central Sweden. Though he couldn't know it at the time, Thure, as he hiked to the shores of Lake Hornborga, was on a journey to the shores of another lake, Lake Koshkonong in the Wisconsin Territory of North America.

When he set off to collect birds at Lake Hornborga, Kumlien would have been equipped much like the young Swedish ornithologist Gustaf Kolthoff, seen here. Kolthoff later became a prominent Swedish writer, taxidermist, and founder of the Biological Museum in Stockholm. PHOTO BY CAROLINA VON KNORRING, VÄSTERGÖTLANDS MUSEUM

The small village of Härlunda on the River Flian dates back to before medieval times. Unlike New England and midwestern villages, in which houses line the roads, Härlunda comprised seven farms centered around an old church. Each irregularly shaped farm with its house and outbuildings included land for crops and pasture and a woodlot. The farms carried old names: Damsgården and Bossgården, Herrtorp, Skattegården, and Ekgården. According to one Skara historian, "The river Flian coming from Lake Hornborga floods the marshes every spring—today as well as two hundred years ago. . . . Then in the summer the marshes dry up giving the farms in Härlunda wonderful hay crops."[5] Though some of the land was stony and meager, the villagers raised potatoes and other vegetables, and grazing was good for dairy herds and beef cattle.

Kumlien's family established a presence in this area in the early nineteenth century. In 1805, Reverend Johan Petter Rhodin and his wife, Elin Landstrom, left nearby Broddetorp Parish and moved into Ekgården—Oak Farm—which had been built in 1785. With the pastor and his wife were their two daughters, eight-year-old Anna Brita and five-year-old Petronella Johanna. According to church records, the two girls were extremely gifted in reading and writing. Little Petronella must have had fine handwriting, because she helped her father, the parish priest, make fair copies of his sermons. Reverend Rhodin, described as a "cheerful gentleman," died in 1812 at the age of seventy. The widow Elin and the two young girls continued to live at Ekgården.[6]

Three years later, a twenty-five-year-old bachelor, "a very wealthy man,"[7] moved to Herrtorp, the estate next to Ekgården. This was Ludwig Kumlien, who came from a mill-owning family in Uppland, north of Stockholm, about two hundred miles from Härlunda. His civil position as Skara regiment scribe and assistant judge provided him the farm Herrtorp as part of his military salary.

The Kumlien family had originated with a miller, Johan Eriksson, from Vik's Mill, Balingsta, in Uppland. The miller's son, Erik Johansson (1692–1768), a member of parliament, fathered two sons who adopted the name Kumlien after their home village, Kumla. The older son, Johan Eriksson Kumlien (1743–1814), employed in Uppland's regiment, had no adult sons,[8] so it was the much younger son, Ludwig Eriksson Kumlien (1756–1839), who passed on the family name. One of his four children was Ludwig Ludwigsson Kumlien (1790–1839), Thure's father.

Ludwig Kumlien was an educated man. He had been a student at Uppsala University and, while there, a member of Fjerdhundra Nation, a fraternity for students from the same part of Sweden.[9] Wealthy and unmarried, Ludwig Kumlien lived for two years as neighbor to the widow Rhodin and her two educated daughters, one of whom wrote poetry.

The year she was seventeen, Petronella Rhodin wrote a sprightly poem grounded in dark thoughts about marriage. She called it "In Praise of Maiden Life."

What a bliss to be unmarried—
Not yet captured in the snare
That robs girls of their joy
And makes their time of happiness so short!
Yes, believe me, girls, the wedding band
Is but a torment to this hand.
Hear my simple song and learn
That maiden life is best.

If a maiden wants to make a visit
she doesn't have to bring
A caravan of little ones,
the seedlings of love,
a grumpy husband who frowns at the sunshine
so there is no joy in paying calls.
No, enjoy the ride!
Your freedom is worth more than that.
. . .

Free as the bird I would like to be!
And give neither my heart nor hand
To any man on this earth.
This is my firm belief and solemn word!
Oh Love, you have many slaves!
You have opened many early graves!
But whether my time be long, be short
As a maiden I would be carried forth.[10]

Many of her girlish fears would come to pass. The same year she wrote her poem decrying marriage, Petronella Johanna Rhodin married Ludwig Kumlien. She moved into her husband's home, Herrtorp, and they lived there for sixteen years until moving back into Ekgården where Petronella had grown up. Ludwig by then had bought three more of the village farms and turned the house at Ekgården into a beautiful manor house with impressive gardens.

Thure Kumlien's painting of his home, Ekgården, was published in a *Milwaukee Journal* article about Kumlien on June 17, 1923. WHI IMAGE ID 147546

On November 9, 1819, when she was nineteen years old, Petronella gave birth to her first child, Thure Ludwig Theodor Kumlien, at Herrtorp. Over the next two decades, Petronella gave birth to at least thirteen more children and was pregnant perhaps five additional times. For most of her adult life, Petronella was expecting a child. She must have given birth almost every year for twenty years. After Thure, the birth years of the Kumlien children (and the death years for two who died quite young) were as follows: Johan August, 1820; Maria Amalia, 1823–1828; Elina Augusta, 1824; Ludwig, 1826; Frithiof, 1827–1827; Carl Rustan, 1828; Oscar Alexander, 1829; Georg Victor, 1831; Amalia Maria Josefina, 1832; Axel Frithiof, 1833; Emilia Johanna Helena, 1834; Maria, 1835; and Knut Hjalmar, 1837.

Most of Petronella and Ludwig's children survived to lead successful and productive lives. The two youngest sons, Axel Frithiof (who was born after his brother Frithiof had died as an infant) and Knut Hjalmar, became prominent architects in Sweden, designing churches, government buildings, and large houses. "The quality of the house Ekgården is remarkable," noted local historian J-G Hemming. "I think that this explains why Thure's brothers were good and famous architects—their splendid manor house inspired them from their childhood, as the marshes and forests with all the birds, animals, and plants inspired Thure."[11] In addition to the architects, one son became a farmer, another a physician, and another the headmaster of a school; one daughter became a teacher. Most of the girls married and bore many children.

Late in his life, Thure Kumlien wrote, "I had one of the best mothers in the world. Her watchful care and influence over me cannot be estimated."[12] Petronella's influence is reflected in the home she made for her family, a home filled with books and music and paintings. Ludwig's money provided a home with servants to help care for the family. And his artistic attention to the farms and buildings made the family's home beautiful as well as secure. "Thure's parents could both draw beautifully," noted descendant Angie Kumlien Main, "as sketches brought by Thure from Sweden testify. They both taught Thure to draw and paint."[13]

This bergfink or brambling was painted by Kumlien in Sweden. COURTESY OF SUSAN BINZEL

Though the family was large, there was enough money for Thure and the other children to receive every advantage. A private tutor was hired to teach Thure to play the flute. When he came to Wisconsin years later as a young man, he brought at least one flute, a transverse one-key flute, as well as the music book prepared for him by this tutor in the 1830s. The handwritten book contained exercises, hymns, and "God Save the King."[14] He played the flute for family and friends and for himself—and sometimes for the birds—all of his life.

When Thure showed an early interest in plants and birds, his parents encouraged him, allowing him to bring into the house the plants he collected and pressed, the bird nests and eggs he found, and birds he collected to skin or mount.

Petronella and Ludwig also provided for their sons one of the finest educations in Europe at Skara Katedralskolan. The Skara Cathedral School, one of Sweden's oldest secondary schools, was founded in 1641.[15] This school, which still exists at a different site, was then in a neo-Gothic building next to Skara Cathedral.[16] In the fall of 1830, eleven-year-old Thure and his ten-year-old brother, Johan August, were sent to Skara Cathedral

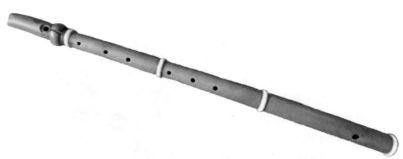

One of two flutes Thure Kumlien brought with him from Sweden to America. COURTESY
OF SUSAN BINZEL

School. Five of their other brothers eventually attended the same school.[17]
Thure attended the primary school from September 1830 until June 1836
and then the gymnasium or secondary school from September 1836 until
June 1839.

In the early nineteenth century, before there was a public school system
in Sweden, each diocese had a primary and secondary school like the one
in Skara for the sons—not yet the daughters—of the nobility and clergy.
With the school being as class-conscious as the society at large, records
were kept about the social class of each student. For example, in 1836, of
the 91 boys in the secondary school and the 271 in the primary school, 11
were listed as nobility, 34 as clergy, 6 as burghers, 14 as farmers' sons, and
26 as commoners.

Like most American school years, Skara's school year ran from early
September to the middle of December and from mid-January to the first
week in June. The boys apparently did not live at the school or at home but
boarded in nearby rented rooms. The primary and secondary schools were
on separate floors of the same building until a growing enrollment forced
a chaotic schedule of alternating use of the classrooms. On August 31, 1833,
while Thure Kumlien was in primary school, a new brick schoolhouse
was opened, for a time alleviating the crowding. But in 1834, the school was
closed for the fall semester when a cholera epidemic sickened eighty
people in Skara, killing forty-five of them. And in the late 1830s when
Thure was in the secondary school, the school experienced a troubled time
with reports of student drunkenness and stones thrown at the headmaster's windows.[18]

In the fall of 1837, a fifteen-year-old student named Otto Vilhelm Olsson, a farmer's son, was expelled—perhaps as a lower-class scapegoat for unrest at the school. Olsson later immigrated to the United States, changing his last name to Akerman. He reappeared in Kumlien's life in an odd coincidence, mentioned in chapter 4 of this book, some years later.

Despite crowding, the move to a different building, illness, and alcohol, the boys at Skara had an opportunity for a fine education, and Kumlien clearly took advantage of that opportunity.

At Skara, he was taught by at least two outstanding teachers: Johan Mathesius (1800–1867) and Johan Wersall (1795–1859). Mathesius, who had a PhD from Uppsala University, "was a teacher warmly interested in his subject matter," which was natural history. To enliven the study of botany, Mathesius took his students on excursions to places such as Kinekulle, a flat-topped mountain on the shore of Lake Vänern, to see plants and birds and insects in varying habitats—stonecrops in the thin soils on the limestone plateau, orchids in the wet meadows, and great oaks in wooded pastures. According to historians of the school, the study of natural history "stood out prominently in Skara under his tutelage."[19] Wersall, also a graduate of Uppsala University, was "a competent, powerful personality" who taught living languages, gymnastics, mathematics, and natural history "dutifully and well."[20] Thure learned from well-trained men who were passionate about natural history.

Though in the 1830s natural history study was not as popular throughout Sweden as it had been in the time of Linnaeus in the previous century, a few of the elite Swedish schools, among them Skara, boasted a "long and successful tradition of teaching natural history."[21] Though zoology was taught to some extent, the study of botany dominated in the natural sciences of the time because botany was then more advanced, systematic, and accessible than other areas of natural science study. Botany, especially in schools, was simply easier to teach. Plants were everywhere, they could be collected without a gun, they were not as messy as animals when dead, and they were easier to preserve and store than animals. The study of botany, focusing on "excursions, the examination of living plants, and the practical preparation of herbarium specimens," was thought to improve the mind as it developed an understanding of plants specifically and of

This sketch of young Thure Kumlien was drawn by one of his brothers in Sweden. Thure brought this with him to North America. THURE L. KUMLIEN PAPERS, WIS MSS MQ, BOX 2

nature in general through the processes of identifying and classifying.[22] In fact, the Skara school had its own botanical garden just outside the city gate.

Useful as a guide and reference to teachers, students, and experienced botanists of the time was Carl Johan Hartman's *Handbok i Skandinaviens Flora* (*Handbook of the Flora of Scandinavia*), published in twelve editions beginning in 1820, including "extensive species descriptions and careful instructions on plant preparation, terminology, and identification."[23] Covering the plants of Sweden, Hartman's *Flora* would have been the handbook everyone owned, the way many people today own Peterson's or

Sibley's field guides, though *Flora* was about more than identification and checklists. It would have been used by Thure and other Skara students for *doing* science, not just *consuming* science.

Skara Cathedral School had one of the most extensive libraries in Sweden, with books on theology, law, medicine, philosophy, mathematics, astronomy, navigation, music, physics, aesthetics, geography, and more. The works in ancient languages—Hebrew, Greek, and Latin, some published as early as the 1550s—supported the classical education the boys received. Books in living languages included literature in Spanish, English, German, Dutch, Danish, Finnish, Icelandic, Norse, and Swedish. English-language books included biographies of Samuel Johnson and Thomas Paine.

The library's natural history collection of more than one hundred works—many of them multivolume—covered mineralogy, fossils, and botany, with fewer on zoology. These included early works of Pietro Mattioli, a botanist and doctor in 1500s Padua, Italy, and works by early and influential Swedish naturalists Olof Rudbeck the Elder and his son, Olof Rudbeck the Younger. The library included the 1603 *Isagoges in Rem Herbarium* in Latin by Adrian van de Spiegel, which explained how to successfully dry herbarium specimens. Once plant specimens were successfully preserved and pressed and mounted on sheets of paper, they could be exchanged, archived, and studied. Thure and his botanical instructors were the recipients of this quiet European "revolution in taxonomy, floristics, and systematics," which had begun several hundred years earlier.[24] Listed also is the second edition of an influential work by Swedish zoologist and archeologist Sven Nilsson (1787–1883), *Skandinavisk Fauna: en Handbok for Jagare och Zoologer* (*Scandinavian Fauna: A Handbook for Hunters and Zoologists*). This 1824 second edition is in Swedish; the first edition was in Latin—the international language of scholarship and science.[25] Often consulted and revised, the detailed descriptive work contained a few hand-colored plates that must have been especially enticing to a young man who loved to hunt, collect, and paint birds. Kumlien took a copy of "Nilsson" with him to North America, though we don't know which nineteenth-century edition.

In the library were at least five works of Carl Linnaeus (1707–1778), including three later volumes of his 1735 world-changing *Systema Naturae* in which he developed his binomial system of taxonomy of plants and animals. Linnaeus's 1746 *Fauna Svecica*—the first published account of Sweden's national fauna—described more than thirteen hundred Swedish mammals, birds, amphibians, fish, insects, and "vermes" or worms.

The Skara library held some twenty thousand volumes; its densely printed 1830 catalog is more than six hundred pages. We can't know which of these volumes Thure read, but we do know what was available to him. Though the natural history section contained many volumes on botany, very few were on birds, reflecting the relative states of the study of botany and ornithology at the time. The innovations in collection and preservation of plant specimens had advanced botany, but the collection and preservation of bird specimens, and therefore the study of birds, had been hampered for decades by problems with insect infestation in collected specimens. In the early 1800s, advances in taxidermy and the preservation of bird skins enabled advances in ornithology.

Many books might have inspired or influenced Thure: perhaps Captain Basil Hall's 1826 *Travels through Chile, Peru, and Mexico* or perhaps the romantic account *Souvenirs d'Italie, d'Angleterre et d'Amerique*, in which Chateaubriand describes the environment of America, Niagara Falls, and the area's Native people. The Western world, with its writings in many languages, was there for Thure on the shelves in the Skara Cathedral School's library. The catalog of books in this library reflects the rich intellectual environment in which Thure spent nearly eight years of his life.[26]

Skara Gymnasium, the secondary school, was important not only for the education Kumlien received there but also for the people he met there. Though older than Kumlien, Gustaf Mellberg (1812–1892) attended Skara from the fall of 1831 to the fall of 1835. Either at Skara or later at Uppsala University, Kumlien and Mellberg became such good friends that they decided to cross the Atlantic together and settled in America near enough to work and socialize together for the rest of their years. Gunnar Wennerberg (1817–1901) attended Skara and later became an important poet and composer at Uppsala University and a close friend of Thure. Magnus "Måns" Cornell (1820–1869) studied at Skara from 1835 to 1839 and went to Uppsala University to study medicine at the same time as Thure. Thure and Måns shared an interest in natural history and later were companions on a collecting trip to Gotland.[27]

As a boy, Kumlien attended Skara Gymnasium for eight years, studying classical languages, sciences, and literature. He had classes in the building pictured here and boarded at a nearby rooming house. VÄSTERGÖTLANDS MUSEUM

In secondary school, Thure received high marks for his studies, which included theology, Latin, Greek, Hebrew, history, geography, living languages (probably French and German, possibly English), philosophy, and mathematics. Continuing his study of natural history at Skara, Thure learned the basics of taxidermy and drew and painted in watercolor birds and animals of Sweden. When he left school to attend the university, he was said to have left behind a little museum of mounted birds and preserved bird skins, of bird nests and bird eggs.[28] Thure gave his collection to the school when he graduated, and later in life, he sent back bird and plant specimens to be donated to the school.

One Sunday evening in 1836, Petronella Kumlien sat down to write to her "dear Thure," who was away at school, likely in the rented room where he and his brother Johan August were boarding. It was seven o'clock. She had spent her Sunday "in the usual way." "Do you know," she asked him, "that in the midst of my household duties or as I am sitting at my needlework, my imagination sometimes conjures up a picture of a dark little uncomfortable room" with her two sons "each in his corner." She wondered how her "poor boys" were getting along and admitted that "many times on these occasions I cry out." Then, perhaps after looking away from the page—thinking and remembering—her tone changed to that of a disciplinarian. She warned Thure: "Let us see now that you are prudent enough not to commit any folly that would only bring pain to yourself and worry to your parents." Though Thure was seventeen and Johan August was sixteen, she told Thure, "Keep a watchful eye on your little brother, who still less than you is able to think and take care of himself." Then Petronella abandoned the role of disciplinarian, which seemed to come less easily to her than that of the expressive, artistic, imaginative woman who amused her children and others with talk. This sentence perhaps best captures Petronella's emotional state at the time she wrote the letter:

> I should very much like to brighten you at the present dark situation[29] with some little jolly tale, or some whimsical remarks, but God knows, the world gives so little occasion for the former, and I, poor thing, am bankrupt of the latter, so I guess it only remains to sing the same old worn-out song like that minister, who made his sermon last the pre-scribed time, by reading it over twice.

In this complex and graceful sentence, even in translation, we can hear the voice of an educated woman who was used to calling upon stories and jokes to brighten the lives of those in her care. Because of current and perhaps continuing trouble, she could not draw upon these methods as she had in the past; she felt powerless. She could only sing "the same old worn-out song."

In the last paragraph, Petronella wrote, "We'll see if I can get a chance to come and see you. . . . I certainly would like to very much. The grand-mothers[30] send their greetings to you with all the power of their old hearts—also Papa and brothers and sisters." She noted the time, eight o'clock. The letter took an hour, spent partly remembering and partly look-ing into the future for her almost-man son, confiding in him. "I bid you good night," she wrote, "by the calling down of God's blessing over you." She signed it, "Your true mother." Thure saved this letter all of his life.

The next year, on March 17, 1837, Petronella safely gave birth to her last child, Knut Hjalmar. That year and on into the spring of 1838, until his mother became ill, all must have seemed right in Thure Kumlien's world. At home, he was loved and comfortable and secure; his interests and talents were recognized and encouraged. Out along the river and in the marshes, in the woods and on the lake, he found a paradise of birds. At school were talented friends; music; and the study of natural science, Greek and Roman classics, and the literature of France, England, Sweden, Germany, Denmark, and Iceland. In the spring of 1838, Thure was seventeen and in his next-to-last year at Skara.

On April 28, 1838, Petronella Kumlien died of cancer, perhaps after a long illness. Thure's world, and that of his eleven younger brothers and sisters, would never be the same. Though his father was still around, as were grandmothers and servants, the family must have been devastated by the loss of the vibrant Petronella. Her fine sensibility and lively humor and her love of music, art, and poetry were part of Thure's identity as well. Her ideas of family life must have been part of what he would carry across the Atlantic with him.

A short time after his wife's death, Ludwig gathered his older children around him—likely seventeen-year-old Thure, sixteen-year-old Johan August, and fourteen-year-old Elina Augusta. Ludwig told his children that he didn't *want* to marry again, but he must have communicated that he needed a wife and they needed a mother. In any case, he asked the children what they thought. The children apparently told their father that it would be the best thing to do.[31] We can imagine their reluctance and resistance, and we can also imagine their fear of living without a mother to bring order to their lives.[32] That same year, Ludwig married Anna Beata Lindblad, the daughter of Abraham Lindblad, a chaplain in the nearby village of Härlunda.

Anna Beata was not a stranger to the Kumlien children, nor to Ludwig. She was a widow and the owner of Kallegarden, one of the farms in Härlunda.[33] She moved to Ekgården to become Ludwig's wife, perhaps bringing domestic order and stability to the grieving Kumliens. Thure returned to school. The first anniversary of Petronella's death came soon after the cranes returned to Lake Hornborga announcing the end of winter. Then Midsommar was marked but likely not celebrated.

On Thursday June 13, 1839, at eight o'clock in the evening, a public commencement ball was held honoring the students who had completed their study at Skara and were going on to the university in Uppsala or Lund. Thure Kumlien was one of those young men.

A few weeks later, typhoid broke out in the village. On July 7, Ludwig Kumlien, Thure's father, died of typhoid. He was forty-nine. Nine days later, Thure's brother Johan August also died of typhoid at the age of eighteen. In just a little over a year, nineteen-year-old Thure Kumlien— and his brothers and sisters—had lost their mother, father, and a brother. Though Thure didn't know it at the time, the chapter of his life with his family in Sweden had ended and a new chapter was about to begin.

2

YOUNG ROMANTICS AND NATURAL HISTORY HEROES

1840–1843

In the autumn of 1840, Thure Kumlien, nearly twenty-one years old, traveled probably by stagecoach service about two hundred miles north to Uppsala University—the school his father had attended before him—in his father's home province of Uppland, north of Stockholm. Despite moving from his wealthy family's home to the Cathedral School to the university, the privileged Thure essentially remained in the same culture—a culture that valued music, art, and poetry, as well as friendship and the observation of nature. What began for him as the diffuse culture of his family—they played music, read and wrote poems, went for walks in the woods—was concentrated and systematized by Uppsala University, where Thure studied literature and natural science. At the end of three years at Uppsala, he had acquired the experience, knowledge, maturity, and supportive friendships to make the two most important decisions of his life.

On the banks of the Fyris River, the university buildings faced the thirteenth-century French Gothic Uppsala Cathedral. Founded in 1477, Uppsala University was the oldest center of higher education in Scandinavia. In rooms within the elegant and stately Gustavianum, the main building where Carl Linnaeus had taught, Kumlien attended lectures on history, literature, and botany. The university library, at that time in the cupola high above the Gustavianum, had been the anatomical theater of the great naturalist Professor Olof Rudbeck the Elder. Kumlien's class would have

Carolina Rediviva, Uppsala University's library building, was completed in 1841 and named for the Academia Carolina library building that had functioned as the school's library since the eighteenth century. This early image was in Kumlien's possession.
WHI IMAGE ID 147686

been among the first to use the beautiful and elegant university library, Carolina Rediviva, completed in 1841.

Thure Kumlien belonged to Västagöta Nation, one of the student "nations" or fraternities traditional in universities of Sweden. With other students from Västergötland, Kumlien celebrated Novichfest, or Newcomers' Feast, and Saint Walpurgis Night. And in the group's building on the Fyris River were held goose dinners and May dinners and spring balls. This was no austere starving-in-a-garret student life. Student life for the primarily upper-middle-class and upper-class young men at Uppsala University was festive, with food and wine, singing and poetry, and musical theater.[1] Thure matriculated at the height of Uppsala's Romantic period, in what was called the "students' century."[2]

A new philosophy of nature, following the rationalism of classicism, swept across Europe in the early 1800s. To the Romantics, nature was suffused with meaning and significance, with a soul, which differed from the materialist view of nature of the pure scientist. "For the romantics, nature was perceived as an organic whole, infused with life and power," explains Swedish historian Bertil Nolin.[3] These ideas influenced scientists of the time, as well as poets and other writers; Romanticism permeated the thinking of some of the most prominent natural scientists of the times, among them Kumlien's professor and friend Elias Fries (1794–1878). Romantic young men who had never done manual labor were attracted not only to adventurous excursions but also to the beauty of toil on the land. This Romantic thinking was eventually carried to Wisconsin by Gustaf Unonius, Kumlien, and other Swedes of the time.

The university in 1840, with its rich social life and its long tradition of the study of natural history, must have felt to Kumlien like both an extension of his previous interests and an oasis after the domestic grief of the previous two years. His family's social status, his economic security,[4] and his fine classical education at Skara would have allowed him to fit in comfortably with other students. A passion for birds and his facility on the flute would have set Kumlien apart but also would have drawn people to him. At Uppsala, this young man, later described as "charming" and "scholarly,"[5] studied among friends, took walks to collect plants and birds, played music, recited poems and sagas, argued about politics, and drank wine and beer at boisterous dinners.

From his Skara days, Thure had known fellow student Gunnar Wennerberg (1817–1901), the writer and composer of the popular students' songs "Gluntarne."[6] Though some were written after he left the university, Wennerberg's well-known songs had their origins in the Romantic student traditions at Uppsala during Kumlien's time there in the "city of eternal youth."[7] The son of a vicar from Västergötland, Wennerberg was a romantic figure—tall and handsome with thick dark hair and a fine baritone singing voice. At the university, he led an elite student society and musical club called the Juvenals. Not long after he left the university, Wennerberg went back to Skara as a teacher and became well known in Sweden for his set of thirty duets for bass and baritone, which "presented an epitome of all that was most unusual and most attractive in the curious university life of Sweden."[8] Wennerberg was described as a "lyrical, ardent, Swedish aristocrat, full of the joy of life and the beauty of it."[9] To be in

Poet and composer Gunnar Wennerberg was the heart of the Romantic movement at Uppsala University during Kumlien's time there. Both Wennerberg and Kumlien belonged to the university's Västagöta Nation student society. Wennerberg is seen here in middle age. WHI IMAGE ID 147393

Wennerberg's circle would have been a heady experience for the quiet Kumlien who did not easily put himself forward. After he left Sweden, Kumlien would never see Wennerberg's dashing figure again, but he sang his friend's graceful, witty songs for the rest of his life at celebrations to remember his student days.

The charismatic poet, historian, and star lecturer Erik Gustaf Geijer (1783–1847) was one of Kumlien and Wennerberg's most popular professors. Geijer was influential in reviving interest in Old Norse literature and Viking mythology. At the same time the Grimm brothers were collecting German folk tales, Geijer was collecting Swedish folk songs and tales. Geijer's work and his lectures gave students, and Swedes in general, access to the ancient heroic Norse literature, folk tales, and songs that became transformed and accessible in Romantic poetry, plays, and songs. Geijer, among others, made the 1840s at Uppsala University "ring with proudly patriotic choruses."[10]

The university students also read many poets, including Johan Ludvig Runeberg (1804–1877), a Finno-Swedish writer of epic poems and of the poem that became the Finnish national anthem. One of Runeberg's books of poems, the 1833 *Lyrical Songs, Idylls and Epigrams*, celebrated the "dogged perseverance" of the life of a farmer and soldier Bonden Paavo (Paavo the Peasant).[11] It may be that Kumlien and the other "soft-handed" students at the university admired Paavo's grit and strength as he fought for his native soil. Runeberg himself admired poetry that used "the real and simple words of the heart."[12]

Though Kumlien most likely would not have known Lund University professor and later bishop Esaias Tegnér (1782–1846), Thure's close friend Gustaf Mellberg did know Tegnér, who spoke at a graduation ceremony in Mellberg's youth.[13] And all of the Uppsala students would have known Tegnér's *Frithiof's Saga*, which tells of the deeds of warriors and lovers in twenty-four songs based on Icelandic sagas. Twelve of the songs were set to music, and *Frithiof's Saga* was translated into all European languages.[14] "For a long time, *Frithiof's Saga* was something of a national epic," notes Swedish historian Bertil Nolin.[15] The immense popularity of the work is understandable in a time when Norse tales and Icelandic sagas were being revived, when poetry was often communal and spoken aloud rather than experienced between a solitary reader and the page. The Swedes who settled near Lake Koshkonong "seemed to know the Swedish poets by heart[,] . . . the great productions of Tegnér, Runeberg, Geijer and a large amount of Sweden's song and story."[16] William Wheeler, a colleague of Kumlien's who knew him near the end of his life, said: "I have heard him repeat with a glow of delight verses from Runeberg and from . . . Tegnér's *Frithiof's Saga*, rendering the wonderful rhythm of the latter with exquisite grace and precision."[17]

The writings of impressionist poet Per Daniel Amadeus Atterbom (1790–1855) combined Kumlien's and other students' interests in both Romantic ideas and the study of nature. In Atterbom's suite of poems, *Blommorna* (*The Flowers*), he gives individual flowers voices and souls. In the Romantics' view, everything in the natural world had a soul. It was "the

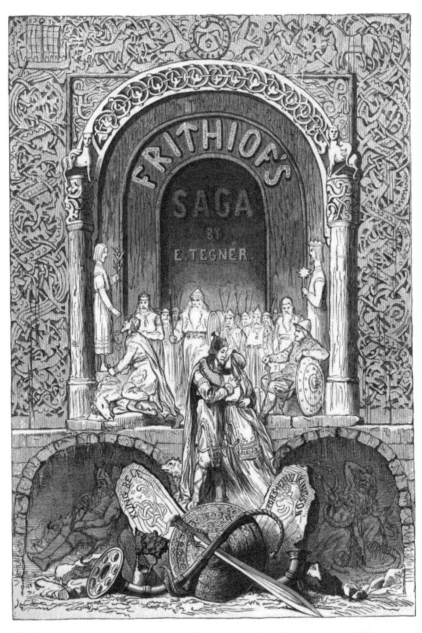

August Malmström's artwork on this title page of *Frithiof's Saga* by Esaias Tegnér gives a sense of the intense Romanticism of the twenty-four canto saga, which Kumlien and many other Swedish immigrants brought with them to North America. PUBLIC DOMAIN, GOOGLE DIGITIZED duty of the poet . . . to describe the world soul as it manifests itself in nature." Atterbom and other Romantics believed that "everything is permeated by something living and everything is one."[18] To the poet, then, the great naturalist Linnaeus was not just a scientist who observed nature but "a priest of nature" who interpreted the natural world. In Atterbom's poem on the twinflower, *Linnaea borealis*, named after Linnaeus, the flowering plant can feel the presence of the long dead Linnaeus:

> What hovers over the woods
> with the song-thrushes final call?
> A venerable shadow, faithful
> To the pleasures of his former home;
> The seer who read me
> And adorned me with his name:
> The priest of nature.[19]

Atterbom wrote more than forty of these dramatic monologues giving voice to Swedish flowers such as the butterfly orchid, snow drop, poppy, and anemone.

Kumlien's Romantic student days at Uppsala remained with him in the friendships he maintained, the songs that he continued to sing and play on his flute, and the poems he recited when family and friends in Wisconsin gathered on Sundays. His attitudes toward adventure, beauty, loss and love, and nature remained Romantic ideas, as far as we can tell, for as long as Kumlien lived.

Perhaps more than anything else, it was Kumlien's knowledge of Romantic literature and ideas that later separated him from the farmers who surrounded him in Wisconsin, with their hard-nosed practicality. His valuing of the unseen, of the soul inside the ordinary and the insignificant, set him apart and made him memorable to those whose minds were open.

Kumlien's life was shaped not only by students and faculty at Uppsala who taught and sang and wrote Romantic literature but also by the illustrious professors and students of natural history who had walked its halls—both during Kumlien's days at the school and many decades earlier. In addition to being a center for Romantic literature and song, the

This painting of a female bearded reedling is one of several artworks featuring Swedish birds that Thure Kumlien brought to Wisconsin in 1843. COURTESY OF SUSAN BINZEL

university had been a center for the study of botany for about one hundred fifty years. The plants in the university's botanical gardens[20] and in its collections of pressed plants had been gathered by renowned botanists, among them Linnaeus and his "apostles." These apostles were nineteen young men sent out by Linnaeus in the last part of the eighteenth century to bring back specimens of plants and animals from far-flung and little-known worlds—specimens Linnaeus would place and name in his taxonomy. Many of these heroic travelers did not survive the dangers of their expeditions. This earlier age of exploration produced plant and animal collectors who were heroes to the Romantic Swedish students leading relatively tame lives in Uppsala in the early 1840s.

Kumlien later wrote of his time at Uppsala that he and "three or four others" (probably Magnus Cornell, Gerhard Von Yhlen, and Carl Gustaf Löwenhielm) were "the only ones who took interest in zoology."[21] Because there was not yet a zoology professorship in Uppsala, those students interested in birds, for instance, had to make up their own course of study by tromping off into the countryside to see what they could see and by occasionally going to Stockholm to study the birds in the collection

As a student at Uppsala University, Kumlien developed a friendship with Professor of Botany Elias Fries, pictured here, and the two developed a correspondence years later in the 1860s. WHI IMAGE ID 147995

assembled by Lund University's zoology professor, Sven Nilsson, at the Swedish Museum of Natural History.

Though Professor Fries lectured primarily on botany, he was knowledgeable about the birds and animals of Sweden. His work at Uppsala and his history as a professor there connected his student naturalists to the work of earlier distinguished Swedish naturalists Olof Rudbeck the Elder and Olof Rudbeck the Younger, as well as of Linnaeus and his apostles. Kumlien and his naturalist friends at Uppsala were particularly inspired by the collecting trips these men had taken to remote parts of Sweden and to distant lands. Though Kumlien, Cornell, Von Yhlen, and Löwenhielm would have read in the sagas about heroic men who fought wars, slayed dragons, and won maidens, these four friends would rather have been heroes in the manner of the Linnaeus apostles, particularly Pehr Kalm.

In the meantime, they studied botany at Uppsala University, attending lectures on mushrooms and vascular plants by the famous Professor Fries, who was at the center of the world as far as the studies of botany and mycology were concerned. He had studied in Copenhagen as a young man, but at the time Kumlien knew him, the young men of Europe were coming to Uppsala to study botany with Fries.

Well known in Sweden and all of Europe for the modern classification of fungi and lichens, Fries was the son of a vicar from the south of Sweden. He studied and taught at Lund University until he was appointed professor of botany at Uppsala in 1834. The tall professor, with his impressive demeanor, wild hair, and lively eyes, was a memorable figure who was then in the prime of his life. Said to be alternately mild and severe, naïve and wise, the intimidating genius nevertheless became a warm friend, inviting Kumlien home to dinner with his wife and many children. Years later, Fries and Kumlien, both remembering their early friendship, corresponded for a number of years, and at least one of Fries's sons warmly remembered Kumlien.[22]

The Uppsala botany students would have known the story of Olof Rudbeck the Younger's lost *Book of Birds*, though Rudbeck was long before their time. Rudbeck (1660–1740), a professor at Uppsala University who was descended from a family of scientists, had written late in the seventeenth century the first book of the birds of Sweden, a book that Rudbeck claimed included "all the birds here in the North—their names, habitats, shapes, distinctive characteristics, and traits."[23] Unfortunately, Rudbeck's text for this work was lost in a great fire in the town of Uppsala in 1702. Though the text of his book was lost, his watercolors were saved. Rudbeck used these watercolors of birds as illustrations in his lectures at Uppsala University—the first ornithological lectures in Sweden. The young Carl Linnaeus, one of his students, took notes. As a tutor for some of Rudbeck's children, Linnaeus had extra time in the Rudbeck household to examine these "splendidly drawn Swedish birds."[24] Later, these drawings became one of the bases for Linnaeus's identification of birds in his important works.

Linnaeus was a towering figure, not only in Sweden but in the history of science. It was Linnaeus who established a method for naming and classifying species that is basically what we use today. In his work *The System of Nature*, a pamphlet first published in 1735, Linnaeus presented his hierarchical system for the classification of nature into kingdoms, classes, orders, genera, and species. He also presented the system of binomial nomenclature—the naming of each species with a two-part Latin term denoting genus and species. Some Uppsala natural history students may have wanted to be university professors like the great men Rudbeck and Linnaeus, Nilsson and Fries, but it doesn't seem that Kumlien or his friends had such desires.

Kumlien, of course, used the taxonomy Linnaeus had created. But he was more influenced by Linnaeus the collector and traveler to Lapland, Öland, and Gotland; Linnaeus the ornithologist author of *Swedish Birds*; Linnaeus the botanist who planned and described the botanical gardens in Uppsala; and Linnaeus the expedition general who "sent out" about nineteen students to collect natural history specimens around the world.[25]

In expeditions approved of or planned by Linnaeus in the last half of the eighteenth century, naturalists left Sweden to report on and collect specimens in China, Egypt, Venezuela, Java, Yemen, Tunisia, Libya, Russia, Japan, and Australia. Christopher Tärnström, the first disciple of Linnaeus, went to China and died there, leaving in Sweden a destitute wife and children. Forced to support the widow, Linnaeus from then on sent out only unmarried men. Eight of the disciples died of disease or injury on their journeys and never returned to Sweden.

Perhaps the most successful and well known of the apostles of Linnaeus was one of his best students, Daniel Solander (1733–1782). Having traveled to England in 1760 to promote Linnaeus's taxonomic system, Solander was hired several years later to catalogue the British Museum's natural history collections, and he was elected fellow of the Royal Society in 1764. Solander also traveled as botanist with James Cook and Joseph Banks on the *Endeavor* on Cook's famous first expedition to Australia and New Zealand in 1768. The expedition preparations and all its gear were described in a letter to Linnaeus:

> No people went to sea better fitted out for the purpose of Natural History, nor more elegantly. They have got a fine library of Natural History: they have all sorts of machines for catching and preserving insects; all kinds of nets, trawls, drags and hooks for coral fishing; they have even

a curious contrivance of a telescope, by which, put into the water, you can see the bottom at a great depth, where it is clear. They have many cases of bottles with ground stoppers, of several sizes, to preserve animals in spirits. They have several sorts of salts to surround the seeds; and wax, both bees wax and that of myrica; besides there are many people whose sole business it is to attend them for this very purpose. They have two painters and draughtsmen, several volunteers who have a tolerable notion of Natural History; in short, Solander assured me this expedition would cost Mr. Banks £10,000.[26]

When the men on the *Endeavor* returned to England, Solander had become the first Swede to sail around the world. Though he stayed in England to work with Joseph Banks and never returned to Sweden, Solander's presence on this famous voyage was known throughout Sweden.

Carl Peter Thunberg (1743–1828) traveled in 1772 as a surgeon with the Dutch East India Company to South Africa, where he collected specimens of flora and fauna for nearly three years. Traveling on to Japan in 1775, Thunberg not only successfully collected flora and fauna of Japan but also exchanged valuable medical knowledge with the Japanese and wrote his masterpiece, *Flora Japonica*, in 1776. Returning to Sweden in 1779 after having been away for about eight years, he found that Linnaeus had been dead for a year. Thunberg's long career as a professor at Uppsala University and the many published works that were based on his notebooks kept the stories and findings of his travels alive in the minds of educated Swedes. Thunberg's botanical collections, about 27,500 sheets of pressed plants, became the foundation of the Uppsala University Botanical Museum, which was accessible to Kumlien and other students.

Pehr Kalm (1716–1779), who entered Uppsala University to study with Linnaeus in 1740, a hundred years before Kumlien attended, may have been the most satisfactory of Linnaeus's apostles, with the possible exception of Thunberg. Kalm's popular journals of his years in North America, published in Sweden in 1753, 1756, and 1761, may have influenced Kumlien's decision to immigrate. In these three volumes, Kalm writes of the first fourteen of the thirty-one months he spent in North America, from September 1748 to February 1751.

Kalm set to work on September 14, 1748, describing the "low, white, sandy" western shore of the Delaware River, the "fine woods of oak, hiccory and firs" he saw on both shores as they sailed up toward Philadelphia. "Here," he writes, "we were delighted in seeing now and then between the woods some farm-houses surrounded with corn-fields, pastures well stocked with cattle, and meadows covered with fine hay." Kalm often describes the smells, as when "the wind brought to us the finest uffluvia of odoriferous plants and flowers, or that of the fresh made hay." He continues, "These agreeable sensations and the fine scenery of nature on this continent, so new to us, continued till it grew quite dark."[27] Actually, with the disagreeable exceptions of mosquitos, flies, and snakes, those "agreeable sensations" continued to fascinate Kalm for two and half years. And they would have done so for even longer if he had had his way.

Kalm's travels through the Delaware Valley, the Hudson River Valley, and the St. Lawrence Valley of Canada are engrossing reading even now, nearly three centuries later. To a young Swede like Kumlien, who loved the natural world, reading the narratives of Kalm, his countryman, that described his travels in North America might have ignited a life-changing dream. We can't know for certain that Kumlien read Kalm, but given the popularity of Kalm's writing and Kumlien's family and schooling, it is difficult to imagine that he did not.

Kalm had been sent to North America to bring back plants that would thrive in Sweden's harsh climate and enhance agriculture there; the interest of Linnaeus and other backers was economic as well as scientific. Yet Kalm, with his economic interest, was the consummate scientist—an omnivorous observer, tirelessly alert, and stimulated by the newness: "I found that I was now come into a new world. Whenever I looked to the ground, I every where found such plants as I had never seen before. When I saw a tree, I was forced to stop, and ask those who accompanied me, how it was called."[28]

In the New World, after he called on Benjamin Franklin in Philadelphia, Kalm went to see John Bartram (1699–1777), a Linnaean botanist who was well known to European collectors and naturalists. Bartram's quiet family life of farming and gardening and observing, interspersed with long, usually solitary collecting journeys, would have intrigued the naturalists in mid-eighteenth-century Uppsala. Through his plant-loving agent in England, Peter Collinson, Bartram's North American plants and seeds were distributed to members of the Royal Society in London and to other botanists in Europe, including to Linnaeus. Bartram wrote, "What-soever whether great or small ugly or hansom sweet or stinking . . . every-thing in the universe in thair [sic] own nature appears beautiful to me."[29] The contrasts between the self-taught Bartram and Kalm, with his univer-sity training from Linnaeus, were apparent to Kalm. Though many of the first North American plants in Europe had come from Bartram, Kalm was critical of Bartram for writing so little. This judgment could later have been leveled against Kumlien, who also wrote little. Though Bartram dis-covered many plants and brought them to the attention of naturalists in North America and Europe, Kalm argued that Bartram was "to be blamed for his negligence; for he did not care to write down his numerous and useful observations."[30] Still, in the mid-1700s, Bartram made many im-portant collecting trips throughout the colonies, as far west as Niagara Falls and as far south as northern Florida. Kalm and others knew that the remarkable Bartram was more an observer and collector than a writer. Only a few extraordinary travelers for science have had the drive, energy, and ability to write as much at the end of exhausting days as did Pehr Kalm.

Bartram was well known in European scientific circles before Kalm met him in Pennsylvania. Botanists of Europe knew Bartram for the plant specimens he sent to Collinson in England, which were then distributed through the Royal Society in London. Though at a slower speed and in a lesser quantity than today, naturalists in the 1730s and 1740s were sharing scientific ideas and observations across oceans. For example, in the spring of 1739 at his farm outside of Philadelphia, Bartram read a ques-tion Linnaeus had posed to his scientist friend James Logan in Philadel-phia. Linnaeus had asked Logan about the functions of the stamen and pistil in the flower of the American sweetgum tree. This question had traveled from Linnaeus in Sweden, through Johan F. Gronovius in Leiden, through Peter Collinson in London, then to Logan in Philadelphia. In-trigued by the question and problem, Bartram gathered sweetgum flowers and wrote up his observations, sending them to Linnaeus through Col-linson and Gronovius. After posing the question in early 1739, Linnaeus had his answer from Bartram by the summer of 1740: both male and female structures were found on the same sprig of the sweetgum tree.[31]

Kalm spent months with Bartram. "When you talk with the man and listen to him," Kalm wrote of his fellow botanist, "you find out that he has been a very keen observer; he is everything: farmer, carpenter, turner, shoemaker, mason, gardener, clergyman, lumberjack and who knows what else. Quite a head on that man!"[32]

When he got back to his home in Finland, Kalm transformed his extensive field notes into little essays on hundreds of subjects. Not only was his writing enjoyed by Linnaeus and other university-based naturalists in Europe, but his journals were also popular—thanks to his objective, accessible, and lively style of writing—throughout first Sweden and then the other countries of Europe once they were translated into English, German, and French. Importantly, Kalm was aware of his primary audience—the people of his home country. His comparisons of North American and Swedish attitudes toward land and crops interested the Swedes of his time, as they do North Americans of our time. For example, noticing the bounty of unguarded pear and apple and peach orchards in the Pennsylvania countryside, Kalm remarked that "the country people in Sweden and Finland guarded their turnips more carefully, than the people here do the most exquisite fruit."[33]

Though primarily a botanist, Kalm also noticed birds. In his essays, he mentioned birds familiar to Swedes and described those he and other Swedes had never seen. "Certain old Swedes told me," he noted, "that in their younger years, as the country was not yet much cultivated, an incredible number of cranes were here every spring; but at present they are not so numerous."[34] Among the North American birds Kalm described were the cardinal, bluebird, mockingbird, and catbird—birds that Kumlien and his friends in the early 1840s had never seen alive. In October 1748, Kalm wrote about the ruby-throated hummingbird, describing "the shining red ring round the neck of a bird not much bigger than a large bumble bee."[35] He recorded how the hummingbirds held their feet in flight, and how they sounded like bees or "like the turning of a little wheel."[36] Kalm did not see the eggs of the hummingbird, as Kumlien would a hundred years later. However, Kumlien may have seen at the Uppsala University Botanical Museum the more than three hundred plant species that Kalm brought back to Sweden or the smaller number of Kalm's specimens at the Swedish Museum of Natural History in Stockholm.

In the winter of 1841–1842 at Uppsala University, Kumlien and his natural history friends attended lectures: Löwenhielm was studying for degrees in philosophy and law, Cornell was studying medicine, and Von Yhlen and Kumlien were studying botany. These four friends were eager to get into wilder country than the land around Uppsala to collect plants and birds. They might have occasionally traveled ten hours by stagecoach or eight hours by boat to the Museum of Natural History in Stockholm to study the collections of birds and mammals there.[37] But one of their friends, Gustaf Unonius, had dramatically left Uppsala for North America the previous summer. Through Unonius's letters, which were published in Swedish newspapers, Kumlien and the other students followed him into Wisconsin, deepening an interest that had been informed by the earlier travels of men such as Pehr Kalm.

3

LETTERS FROM AMERICA AND A TRIP TO GOTLAND ISLAND

1841–1842

In early January 1842, Thure Kumlien and his naturalist friends at Uppsala read in *Aftonbladet* their friend Gustaf Unonius's description of the beautiful autumn in North America. Unonius's portrayal of the forests, rich valleys, clear rivers, and lakes of the Wisconsin Territory must have set the naturalists dreaming about studying the plants and birds of that unspoiled and unstudied new world. One of these friends, Carl Gustaf Löwenhielm, a bird collector and student of philosophy and law, had the money to make one of Kumlien's collecting dreams come true, while Unonius had been the first of the Uppsala friends to act upon his romantic dreams, influencing not only Kumlien but many other Swedes.

In the early spring of 1841, thirty-year-old Unonius took stock of his life. He looked around his room in Uppsala, which was austere even for a student's living space. What he saw was a spiderweb-covered bookcase with a meager collection of books and a writing table under the window holding an inkstand and piles of bundled manuscripts. In earlier years, the piled papers had been his student papers and manuscripts of his poetic writings. But now this pile of papers was, depressingly, not his own work; it was the government land office documents he was hired to copy. Unonius was no longer a student or an aspiring writer; he was a clerk with no other future that he could see. This was, to him, "a sad chapter in . . . a young man's life history."[1]

Gustaf Unonius was an important figure not only in the life of Thure Kumlien, but also in the lives of other Swedes who immigrated to North America after reading his descriptions of Wisconsin in the Stockholm newspaper *Aftonbladet*. WHI IMAGE ID 69868

Though Unonius was nine years older than Kumlien, the two had become friends at Uppsala when Kumlien was a student. Unonius, born in 1810 in Finland to educated parents, had graduated with a law degree from Uppsala University in 1830. As a student, he wrote poems, plays, and a novel and "played an active role in the pulsating student life."[2] In 1833, when the university was temporarily shut down because of cholera, Unonius had performed volunteer medical work, apparently liked it, then tried to go back to the university as a medical student. Though his grades were not good enough for him to be admitted to study medicine, he wanted to stay on in the vibrant university town, so he took a clerkship in the provincial seat of Uppsala. It was at this time, when he was a clerk in the land office of Uppland, that he and Kumlien must have met and, sharing an interest in music and literature, struck up a friendship.[3] In the spring of 1841, in this time of "bitter disappointment," Unonius thought about America: its "rich soil and industrial advancement" offered him "a home, a means of livelihood, and an independent life."[4] Unonius, like Kumlien and many of the Romantic upper-class youths of the time, thought that labor, "if only honest, was no disgrace."[5] Unonius was in love with Margareta Charlotta Öhrströmer, and he planned to marry her, but how would he support himself and his wife on the pay of a clerk? What sort of future awaited them in Sweden?

America drew him: "Her fabulous birth and history had excited our wonder from our earliest school years," he wrote, "when we learned to point out its position on the map." As a schoolboy, Unonius—and likely Kumlien—had seen America as a "cradle of true civil liberty, equality, and of such new social ideals as are destined to bring happiness to mankind." At age thirty, what he referred to as "the springtime of my life," Unonius believed he would find in America what would take years to attain in Sweden. In a sentence that seems to refer to a well-loved poem, Unonius summarized what he hoped to find in America: "A faithful, loving *heart* was mine already; the *cottage*, in which the youth and the poet find their dream of happiness fulfilled, was already growing in the primeval forests and only awaited my hands to be built."[6]

In later years, between 1846 and 1930, about one and a quarter million Swedes would immigrate to North America.[7] Only a handful left before that time. In 1840, when Unonius was imagining the cottage in the forest, there were probably fewer than a dozen Swedes in Wisconsin, perhaps only six—Carl Friman and his five sons on a farm near Kenosha. So, Unonius and then Kumlien were not on a well-traveled path in 1841 and 1843. Later, Swedes such as the Bishop Hill colonists left Northern Europe for religious reasons; they felt they could not exist within the framework of the Swedish state church. During this time, religious revival movements began to arise in reaction to the social and political discontent and to the "dry formalism" of the Lutheran state church. An increasing population resulting from "peace, vaccine, and potatoes"[8] caused farmland shortages, sending the immigrants (primarily farmers) across the ocean in search of cheap land. But pressures on agrarian life were not Unonius's or Kumlien's reasons for leaving Sweden.

The Sweden of the 1840s was very unlike the egalitarian, progressive society of today; Sweden today is known for its high standard of living and low level of income inequality. The Sweden that Unonius and Kumlien grew up in was stratified into four estates or classes: nobility, clergy, burghers, and peasantry. Movement among these rigid classes was rare. The two men would have known members of the nobility, the clergy, and the peasantry, but they would have expected to be members of the burgher class for all of their lives. Life in Sweden at the time was pervaded by social distinction: "A person's station in life could be immediately identified by his occupation, manner of speech, customary form of dress, and exact degree of deference or condescension with which he was treated by others

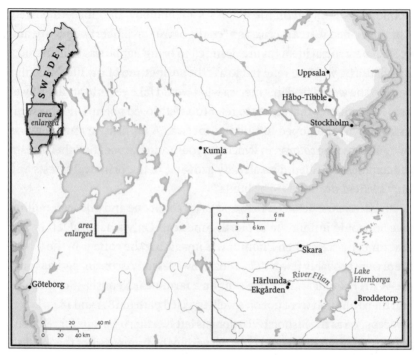

As a boy, Thure Kumlien lived with his family at Ekgården near Lake Hornborga and went to school in Skara. His father's family came from Kumla and his wife's family from Håbo-Tibble. He attended university at Uppsala. MAP BY MAPPING SPECIALISTS

in the social scale."[9] For example, there existed four proper ways to address a married woman and four proper ways to address an unmarried woman in Sweden; in North America, there were two ways: Miss or Mrs.

In the spring of 1841, though Gustaf Unonius was a clerk in a dead-end job, the well-educated and adventurous Margareta Öhrströmer agreed to marry him. And she agreed to leave Sweden and go with him to North America, though this was considered at the time "something extremely strange." They were married on April 26, 1841. In his memoirs, Unonius described two weeks of a honeymoon and a social whirl with "each day filled with invitations from one or another of the numerous families in Uppsala with whom for years I had enjoyed social intercourse and who were now vying with each other to show us the warm sympathy that accompanied their good wishes."[10] As one of Unonius's friends, Kumlien likely attended some of these gatherings.

On May 11, 1841, as the newlyweds left Uppsala for the Port of Gävle, beginning their journey to North America, a retinue of twenty carriages escorted the party, which included Margareta's maid, Christine Södergren; two friends in their early twenties, Ivar Hagberg and Carl Groth; and the dog Fille running beside the wagon. "The streets were thronged and the windows were darkened by people eager to catch a glimpse of the unusual sight and to wave goodbye," Unonius recalled.[11] In a reference to one of the songs written by their friend Gunnar Wennerberg, Unonius likened the procession of singing students to "knightly guards of light."[12]

Kumlien—as well as his longtime friend from Skara Gymnasium and Uppsala University, Gustaf Mellberg—would have been among the company who pressed Unonius's hands in farewell. He would have been in the waving and singing retinue of carriages accompanying Unonius and his party as they left Uppsala. The last song to be sung "from the city of youth," Unonius remembered, was "the beautiful idyll of Valerius: 'I hail thee, thou peaceful streamer! / Come flag-adorned ship for the dreamer / Rock safely the lovers to sleep!'"—another song written by Wennerberg, who might have been there as well.[13] Then, as the students were left behind, Unonius wrote, "The tones of the song died away, and we were left to ourselves to continue in each other's company 'our journey to the port at the end of the world.'"[14] Kumlien would soon follow Unonius to that same port at the end of the world.

The excitement caused by the emigration of the Unonius party was, in large part, a result of the rarity of Swedes leaving for North America. In that year, only two ships left the near Port of Gävle for New York, and ten others left for New York from Stockholm and Göteborg. Twenty-five Swedes arrived in New York that year from Swedish ports; while many of those were relatives of ship captains or sailors, five of them were the Unonius party. That memorable day in Uppsala surely made an impression on Kumlien.

Unonius and his party arrived in New York on September 10, 1841, then traveled by steamboat and canal boat and steamboat again around Michigan, arriving near the end of September in Milwaukee, "a little five-year-old city with 3500 inhabitants."[15] From Milwaukee, on October 13, 1841, Unonius wrote a letter to be published in the important Swedish newspaper *Aftonbladet*. This informative and influential six-thousand-word letter, published in early January 1842, was read eagerly by all of those left

behind in Uppsala, including Kumlien. And it may be that, to his friends
Kumlien and Mellberg, Unonius wrote encouraging personal letters describing prospects in Wisconsin.[16]

While Margareta and Christine stayed in Milwaukee, Unonius, Groth,
and Hagberg traveled on foot about thirty miles west, generally following
an old American Indian trail, which would soon become the Watertown
Plank Road, into the "interior." On the shore of Pine Lake, in what is now
called Chenequa, they selected land for their home.[17] On October 12, the
three men began to cut the timber for their log house, which they hoped
to move into two weeks later. "In the meantime," Unonius wrote in a letter
to be published in *Aftonbladet*, "we shall lodge in a hut built of the boards
that later will form the floor and the roof of our cottage."[18] Just as he had
dreamed of doing less than a year before, Unonius was building a cottage
in the primeval forest. He continued in his letter:

> Wisconsin Territory, where we have settled, is for the present con-
> sidered to be the most favorable region in the United States for colo-
> nization. The country is beautiful, adorned with oak woods and
> prairies broken by rivers and lakes swarming with fish; and in addi-
> tion it is one of the healthiest areas in America. . . . We have taken
> 160 acres in a section located twenty-nine miles from Milwaukee.
> We are the first settlers here. We have deliberately chosen this section
> in the hope that a number of our countrymen will join us. The loca-
> tion is one of the most beautiful imaginable. On the map in the land
> office it has already been designated "New Upsala."[19]

This letter consisted primarily of advice to those who might want to follow
him from Sweden to Wisconsin. To those back in Uppsala (sometimes
spelled Upsala by Swedes), Unonius seemed to have the reliable information that could turn dreams into plans. He outlined a path with steps that
could lead from Sweden to America and relayed to his friends what he had
found there: Wisconsin was a beautiful place not unlike Sweden, but where
one might buy cheap land near a lake, with trees for building a cabin and
with fish in the nearby waters.

Unonius, a well-read man and a skilled writer, took pains in this letter
and his next one to make himself a reliable narrator, comparing various

works on travel to America that he and like-minded Swedes would have read. He advised that Alexis de Tocqueville's 1835 *Democracy in America* was more reliable than Frances Trollope's 1832 *Domestic Manners of the Americans* and that the 1830s accounts of North America by the Swedes C. D. Arfwedson and Karl August Gosselman—both of which could be found in the Skara Gymnasium library—"painted too bright a picture."[20]

In this lengthy letter, Unonius told would-be immigrants how much land cost and how to avoid crooked land agents by buying only from the government land office and, even then, hiring an English-speaking lawyer to look over the documents. He communicated, in pell-mell order, how to build a log house, what tools and clothing to bring from Sweden, how much money everything cost, how and where to book canal and steamship transportation to Milwaukee from New York. He warned against various forms of thievery, he tempted with descriptions of beautiful land, he clarified and he qualified and he warned:

> Even those in the homeland who, like myself, are doomed to a future of indebtedness and dependency, either thronging the civil service or bowing in the antechambers or as farmers burdened with taxes and not owning their plots of land or as narrowly-defined craft workers lacking opportunity to work—all should carefully examine and deliberate before they bid farewell to a country which, no matter what our fate may be, will always be remembered with regret and love, if for no other reason than the tender bonds of relationship and friendship which unites us with the fatherland.[21]

Several times in this first letter, Unonius made the point that it was easier if a group of people came to settle in America together, one of them speaking English, in order to share expenses and labor.

In January 1842, when Unonius's first letter was published in *Aftonbladet*, Kumlien was at Uppsala University studying with Elias Fries. That April, the first of the Friman America letters from Wisconsin were published in *Aftonbladet*, reiterating and intensifying what Unonius had said about America. The widower Carl Friman, who had been in the same regiment in Skara as Kumlien's father, had left Sweden for America with his five sons in 1838. Settling near Southport, what is now Kenosha,

Wisconsin, the six of them for a year lived the settlers' life of clearing and planting, building a cabin and outbuildings, and splitting rails for fences. When Carl and his youngest son returned to Sweden due to illness, the other four sons—the oldest one eighteen—remained and carried on, writing letters home to Sweden and, over time, becoming successful American citizens. The roughhewn and youthful letters of the Friman brothers were written in a different tone than the detailed and literary letters of Unonius; Hurry, the brothers said again and again, the good land is going fast. And a subtext was: if these children can make it on their own in America, any Swede can.[22]

Then, in late May and early June 1842, more letters from Unonius were published in *Aftonbladet*. These contained more detailed and even more beautiful descriptions of Wisconsin lands—the prairies, the savannas around the lakes, the lakes themselves. He wrote, "Our house is in an oak clearing hard by the shore of a lake. Before us, within gunshot, lies a beautiful peninsula overgrown with oak, American pine, and fir trees."[23] He described lakes full of fish and covered with ducks and geese. In the forest were deer and wild turkeys, and birds Unonius identified as pheasants and prairie-chickens.[24] He told of fertile soils and a healthful climate, a beautiful and productive country watered by many lakes and rivers, prairie and oak openings. Of the oak openings, which Unonius saw as the most common land in Wisconsin, he wrote: "Hills and dales overgrown with oak, hickory, walnut, and other trees . . . impress me as the most beautiful landscape in the West. They present a beautiful panorama of green hills and grassy vales. You see straight, leafy oaks, grouped here and there like lanes in an English park, and between each row is an open grassy space bedecked with thousands of flowers. They often form a border around crystal-clear inland lakes."[25]

Also reading the Unonius letters was Kumlien's bird-loving friend Carl Gustaf Löwenhielm (1820–1906), the son of a wealthy clergyman, born February 1, 1820, in Uppsala just a few months after Kumlien. Ten days after his birth, Löwenhielm's mother died from complications during childbirth. When he was two, his father married a woman named Betty Ehrencrona. A son, Gösta, was born in 1823. In 1826, when Löwenhielm was six, his father died, leaving his widow to raise the two boys. Betty moved the family to Stockholm where she became the first headmaster of

Wallinska School for Girls. The Löwenhielms were related to Uppsala professor and Romantic poet Erik Gustaf Geijer.

Löwenhielm had been counting on graduating from Uppsala in the spring of 1842, but some unnamed misfortune delayed both his graduation and a trip to the north of Sweden to study natural history and collect specimens. Löwenhielm had been planning to take Kumlien with him, paying his friend's way because Kumlien had previously been useful to him in identifying and collecting birds. But this trip to the north never happened.

Partly to allay Kumlien's disappointment and partly to get the plants and birds he knew Kumlien would collect, Löwenhielm offered Kumlien a collecting trip to Gotland, the largest island in the Baltic Sea. Gotland, about the size of Long Island off New York, is about sixty miles east of the coast of south-central Sweden. According to Löwenhielm, Kumlien hesitated at first, not sure that he wanted to go alone. But then their friend Magnus—called Måns—Cornell offered to accompany Kumlien to Gotland.[26]

Kumlien had known Cornell since they were boys. Cornell was born in Skara one year after Kumlien, and his father was an inspector at a manufacturing plant. A student at Skara and then at Uppsala University at the same time as Kumlien, Cornell studied medicine at Uppsala, which included the study of botany. The friendships of Cornell and Kumlien and Löwenhielm were likely based in their shared interests in botany and zoology.

So Kumlien's collecting trip to Gotland came about as a consolation for the canceled trip to northern Sweden, with Löwenhielm paying Kumlien's expenses and perhaps also those of their friend Cornell. Löwenhielm may have paid because Kumlien didn't have the money, or he may have paid because he wanted to "own" the results—the bird skins, the plant specimens, and the field notes.

With the open-handedness of his friend Löwenhielm and the company of his friend Cornell, and against the backdrop of his friend Unonius's more dramatic journey, Kumlien took a step that would, in unexpected ways, lead to his journey to North America.

Because of their university class schedules, neither Kumlien nor Cornell could leave for Gotland until the end of May, and they had to be back in August. This timeframe was not ideal because they would miss the spring and fall bird migration seasons on Gotland.

From Uppsala, Löwenhielm saw the two off on May 29, 1842.[27] Almost exactly 101 years earlier, on May 15, 1741, Linnaeus had set off from Uppsala on a well-documented collecting journey to the island of Gotland with six of his students. And twenty years before Kumlien and Cornell's trip, their botany professor, Elias Fries, had also traveled to Gotland to collect plants and birds. Kumlien would remember the date of his trip in his journal two years later, on May 29, 1844, as he worked his field near Lake Koshkonong: "Two years ago today since Mans and I went to Gottland."[28]

Kumlien and Cornell, after stopping for a few days in Stockholm to buy ammunition—black powder, wadding, and shot—and to collect information, took the steamship Gotland from Stockholm to Västervik, the city on the east coast of Sweden where they would embark for the island. While waiting for the boat, they spent a day looking for interesting plants and collected, among others, the yellow-flowering common gorse and the yellow wood anemone.[29]

On June 3, the steamship Actif took them from Västervik to Visby on the west coast of the island of Gotland.[30] The medieval walled city of Visby, constructed of the region's native limestone, had been a near ruin in the time of Linnaeus. The crumbling city wall and churches had reminded Linnaeus of images he'd seen of the ruins of Rome. Today it is a major tourist site, but in Kumlien's time, Visby was being restored and was just entering a time of renewed trading and tourism. With more accommodations than in the time of Linnaeus and before the tourist crowds of today, Gotland was, in 1842, a quiet fishing and agricultural island.

Kumlien and Cornell arrived on Gotland in early June. Because the distances between the places they hoped to visit on the flat island were not great, they could easily walk or hire horses. Ahead of the young men was a whole summer—a summer of long days with moderate temperatures and constant sea breezes, delicious days to spend observing the island's nesting birds and the summer-blooming orchids. Though the trip to Gotland lacked the romance of an arduous foreign journey, Gotland was at least different from the hills and farmlands and forests of the mainland Sweden that they knew.

Gotland is essentially a flat plateau of limestone, tilted slightly to the southeast, with cliffs and bluffs on the western side. Though there are no

When Kumlien and Måns Cornell traveled to Visby, Gotland, they likely explored the medieval church ruins seen in this watercolor painted by Lars Cedergren, ca. 1816–1830. In 1842, the restoration of the medieval city, including the church of Saint Lars on the right, had just begun. PAINTING PHOTOGRAPHED BY LENNART STRÖMBERG, SWEDISH NATIONAL HERITAGE BOARD

rivers and only a few streams on Gotland, there are shallow lakes and springs and marshes. With little topsoil and few forests, the bird lovers found rich bird life on the shores and limestone cliffs and in the marshes, which were even then being drained for agriculture. Kumlien had seen some draining of the bird-rich marshes surrounding the Lake Hornborga of his childhood, and he would see much more at Lake Koshkonong in Wisconsin. With thin topsoil, Gotland had never been a heavily forested place, and even by Linnaeus's day, the need for wood to fuel lime kilns had despoiled the limited forests. But botanists did not go to Gotland for forests; they went for the plants of the alvar.

An alvar is a relatively rare and specialized biological environment in which thin soils on exposed, flat limestone bedrock support sparse prairielike plants, lichens, and mosses with few or stunted shrubs and trees. Because alvars are often flooded in the spring and then very dry in the summer, they are stressed habitats supporting rare plants and animals. Orchids, which many assume to be only tropical plants, are abundant

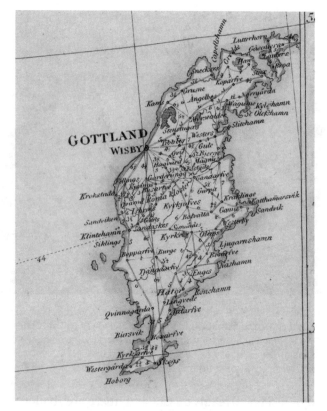

Gotland (or Gottland according to this Swedish map from 1843)
sits sixty miles off the east coast of Sweden in the Baltic Sea.
Kumlien and Måns Cornell landed at Visby (or Wisby) in June 1842
and, for much of that summer, traveled up and down the east
side of the island. They also visited Little and Big Karlso Islands,
which are not labeled, but can be seen on this map off the southwest coast of Gotland. LIBRARY OF CONGRESS, GEOGRAPHY AND MAP
DIVISION

on the alvars of Gotland and the nearby island of Öland and on the alvars
around the Great Lakes in North America.

Once they were on Gotland, perhaps conscious of the money it was
costing Löwenhielm, Kumlien reported that he and Cornell immediately
began collecting,[31] though they must have walked around Visby's narrow
cobblestone streets admiring the impressive medieval wall and towers,
the ruins of churches, and the old cathedral. While in Visby, Kumlien and
Cornell stayed at an inn owned by the forty-two-year-old Carl Johan

Chasseur, the Danish vice consul on Gotland. Chasseur, who was born on the island, was himself a collector of birds and a source of information about where to go to collect and what they would find.[32] They seem to have found a kindred soul in Chasseur.

The first day, June 5, Kumlien and Cornell followed in the footsteps of Linnaeus by leaving Visby for Martebo, a tiny village with an impressive Gothic church about nine miles north.[33] They may have walked, but with the equipment they were carrying—guns and ammunition, food, extra clothing, vascula to put birds and plants in, and collecting materials—they probably hired horses or a horse and cart. On the white limestone roads nearing Martebo, they would have seen the large church rise above the flat land shimmering in the summer heat. Once at Martebo, they must have rented a room in the little village or at a nearby farm. Perhaps they walked into the cool church built in the middle of the fourteenth century to see its carvings of the life of Christ based on the faces, clothing, and tools of the peasant men and women of Gotland.

Kumlien and Cornell spent several weeks in the area of Martebo, wading into the Martebo bog to collect plant specimens and to shoot the birds that lived in the marsh. "The bog," Kumlien wrote, "is hard to walk in, sometimes one sinks as deep as one's crotch. Water washed in and out of our shoes."[34] They noted that they were collecting diligently in this area and at nearby Lummelunda until about the end of June.[35]

The birds they shot they processed immediately by skinning them, perhaps in the field or at least at the end of the day. They treated the skins with powdered arsenic, stuffed them temporarily with moss or dry leaves or paper, recorded them as male or female in a field book, wrapped them in paper, and put them into collecting bags. No work and no effort, Löwenhielm wrote later, had been spared to collect the birds.[36]

Wherever Kumlien and Cornell walked on wet ground, the long-necked, potbellied shorebirds called ruffs—the most populous species on the entire island—ran in the grass in front of them, probing the mud for insects. In the Martebo bog, the men saw and killed many ruffs, primarily females. While they saw groups of males in the distance, they couldn't get close enough to collect many.

In the marsh were wading birds. Kumlien wrote that the wood sandpiper was "not as common as at home."[37] What he identified as *Totanus*

This 1850s photograph of the Lummelunda Church in Gotland, Sweden, shows the white limestone road that Kumlien and Måns Cornell would have walked upon and the fences found all over Gotland, called *bandtun*, which are made of diagonally placed boards tied with withes to upright poles. COURTESY OF MONOVISIONS

hypoleucos, the common sandpiper, was called the *Tringa hypoleucos* by Linnaeus in the 1758 edition of *Systema Naturae* [and is now called *Actitus hypoleucos*]. Kumlien and Cornell saw a large number of the marbled brown, wary, and noisily piping redshanks. In the meadows were woodcocks and lapwings and the chiffchaff, a warbler intensely singing its sweet repetitive song.

The black-tailed godwit, which they saw only at Martebo, was their most prized ornithological specimen. They reported that the godwits were so protective of their eggs and young that finding a nest was impossible. "Still we have not been able to find out if the [godwits] have nests or chicks," Kumlien wrote. "Sometimes they are completely frantic and scream, as if one of them has decided to kill the other by screaming."[38] And then when they were shot, they were difficult to find on the ground. Of ten godwits brought down, the men could find and retrieve only one. By June 22, Kumlien, not surprisingly, reported that the godwits "in this place begin to become sparse, and much shyer than before."[39]

By this time, just over two weeks into their trip, Kumlien and Cornell had used up the shot, powder, and wadding they had bought in Stockholm. Used to shooting birds in trees or on the ground, they had to try to shoot birds in flight in the open marshes, which was much more difficult.

Though ammunition was expensive, they returned to Visby and bought it anyway, sparing no money or effort to collect birds.

They captured birds in nets as well, capturing two guillemots as well as some common terns, common gulls, and oystercatchers near Martebo. As they left the area and returned to Visby by road, they missed shooting the black-headed gull. In the parish of Lummelunda, they would have seen the disappearing stream that Linnaeus described as "wondrous"—a stream that began in the Martebo marsh and ran underground through the limestone "below hills and valleys" emerging at the edge of the bluff on the western side of the island.[40] And perhaps they saw the magnificent green-winged orchid. In the marshland, Linnaeus had noted, "the great fen-sedge, which was used to thatch roofs, grew everywhere as in a cultivated field of grain."[41]

In the account by Löwenhielm, which we have as the only report of this Gotland trip, plants are scarcely mentioned. Yet Löwenhielm stated at the beginning of this report that Kumlien and Cornell could not go "during the birding periods of spring and fall, so the botanical harvest was what we were most hopeful about."[42] Kumlien, who is rightly most known for his knowledge of birds, was also a passionate and well-trained botanist. He had studied botany at Skara and at Uppsala with Elias Fries, one of the greatest European botanists. Later in his life, he continued to study and collect plants in Wisconsin. Kumlien would have noticed and collected the rare and beautiful plants of the marshes and the alvar on Gotland, but we do not have his original field book, which Löwenhielm was working from to create his report. Though it is possible that Kumlien didn't mention plants in his field notes, it is more likely that Löwenhielm, whose primary interest was birds, selected material from Kumlien's report to write a paper about the birds collected on Gotland. As he prepared the draft of that paper, he may have left out the plants that Kumlien likely noted. In a letter decades later, Kumlien wrote of the "many rare specimens, both of plants and birds,"[43] that he had found on Gotland Island.

Back in Visby in late June, Kumlien and Cornell made day trips out to the marshes and the seashore. They also spent time with Chasseur, whose collection of birds from Gotland they believed to be "rather beautiful and the birds well preserved."[44] They went to the Visby Gymnasium to see the bird collection there, some of which had been donated by Chasseur the previous year.

On July 5, Kumlien and Cornell found the rare *Sterna nigra*, the black tern [*Chlidonias niger* in modern usage], south of Visby at Hejde bog in central Gotland.[45] Harvesting the black tern was especially significant because the men knew it had "not been captured since Professor Fries took one down 20 years ago."[46] These small and graceful silver-backed terns with their black heads, necks, and bellies would have been tending their floating nests in a breeding colony of up to a hundred squeaking birds. The restless little birds, like flycatchers or nighthawks, dart over the bog after moths, as light and buoyant as butterflies.[47]

At Hejde bog, they also collected the spotted crake, a small brown bird with yellow bill and green legs. Here they found, probably nesting on the ground in the bog, the little gull with its neat black hood, thin bill, and red legs. In the water, they collected "the good insect,"[48] *Hydrophilus piceus* or the great silver water beetle, a large omnivorous beetle that can be almost two inches long.

After several days at Hejde bog, Kumlien and Cornell traveled back west to the coast where they made arrangements to sail several miles out, first to the island of Lilla Karlsön (Little Karlso) and then to Stora Karlsön (Big Karlso). Hunting was usually not allowed on these islands, but Kumlien and Cornell had received hunting permits "from the landowners—a bailiff and a priest."[49]

Little and Big Karlso are cliff-sided limestone plateaus in the Baltic Sea with large breeding colonies of cormorants, guillemots, murres, razorbills, and gulls. The smaller island has an area of about six-tenths of a square mile, and the larger island is nearly a square mile. On both windswept, sheep-cropped islands, the white limestone is covered with either golden lichens or thin soil with short grasses. Kumlien and Cornell were probably too late for the purple and white orchids that bloom in June on the alvar, but there were surely many other plants blooming alongside delicate ferns in sheltered cracks in the rocks. When they weren't hunting birds, the two men could have wandered around looking into caves, scrambled up and down cliffs, and gone swimming on a white beach. Kumlien and Cornell took food with them for two days, but it seems they stayed a week or more.

On the Karlsos, they found birds in such large numbers that it was "as if the birds were bees swarming around a hive."[50] The number of birds, then as now, must have been almost overwhelming. And Kumlien reported

This 1847 watercolor, painted by Gotland artist Pehr Arvid Säve, depicts a cottage and *bandtun* fences in Träkumla, Gotland—a scene like many Kumlien and Måns Cornell viewed on their trip in 1842. UPPSALA UNIVERSITY LIBRARY

that "the lesser black-backed gulls, in countless multitudes with common gulls and great black-backed gulls, crossed the air incessantly and their penetrating screams . . . woke us up in the mornings."[51]

Though it is not in their report, Kumlien and Cornell were in time to see one of the marvelous spectacles of these cliff bird colonies. The newborn guillemot chicks—really just balls of fat and fluff—are coaxed by their parents to jump from the ledges of the cliffs into the sea below.

Though the men were constantly loading and shooting at the common guillemots,[52] razorbills, and black guillemots, Kumlien wrote that the birds were not frightened by the musket reports but flew fast and straight in a whirring flight. Most of their shots missed until Kumlien and Cornell learned to "lead" the birds; they had to "aim in front of the birds instead of at their bodies."[53] They shot ten razorbills and three common guillemots, also called murre. They reported that they saw common shelducks, which they had also seen on the beach in Fröjel and Västergarn in large flocks, but these were so shy or flew so high and fast that they were "impossible to reach."[54] Still, they did get several of the shy Eurasian rock pipits. On

Stora Karlsön, they shot five razorbills, one common guillemot, a male and a female Eurasian oystercatcher, and seven lesser black-backed gulls—"damned beautiful gulls."[55] Another hunting party there at the same time shot about fifty birds. Of these, several were given to the less experienced, or less successful, Kumlien and Cornell. Common guillemots were found in large numbers on both islands but not to the same extent as razorbills.

Once, when they took a down-covered razorbill chick from its nest, the young bird "nipped soundly and screamed," startling them with its "hot temper."[56] And on another occasion, when Kumlien reached with an oar to collect a gunshot-wounded razorbill in shallow water, the retrieved bird bit him so hard that he bled.[57]

On Stora Karlsö, Kumlien and Cornell also found a rare, pale pink, five-petalled rose similar to the prairie rose, which Kumlien would later see in Wisconsin. Though Löwenhielm did not mention this find in his manuscript, Kumlien's discovery of the rose was referred to in a botanical article on Gotland plant topography and plant geography more than fifty years later.[58]

After their intense bird hunting on Big and Little Karlso, the two men were back on the main island, at Vamlingbo, the very southern tip of Gotland, on July 14. In Muskemyr, they searched in vain for the little gull. But they did get five black-headed gulls and a female tufted duck, which they were "surprised to find, but saw no more of that bird."[59] They also shot seven mature horned grebes and four down-covered grebe chicks. Talking to farmers, they learned that rooks were not rare in the area and sometimes appeared in large flocks. They found black-tailed godwits and the Eurasian hoopoe, though they didn't report that they shot any. Kumlien's observations of the little tern conflicted with Chasseur's: "Chasseur's information that the little tern was to be found in Muskemyr was not confirmed and I think that by no means is it common here."[60]

On July 18, only a couple of miles from where the pied avocet was commonly found, the men were low on money and out of ammunition: "So in disappointment we left *Recurvirostra* [the pied avocet] in peace—for another time!" The next day, "on Sara's Day, July 19, we went hunting, wading as usual, in the rain, for the black tern."[61] That day, at Storträsk at Roma, they shot six more black tern. Because they had decided to leave Visby on

July 23, they wrote, "The pied avocet can thank the lack of money, time, and ammunition for their lives."[62]

The last page of Löwenhielm's draft account of the trip lists the monetary value of the collected specimens on the Gotland trip at 71.46 riksdaler banco, or in today's exchange, about nine hundred dollars. But the birds that were killed by Kumlien and Cornell had more than monetary value. "As bird collections became established and grew," explains curator Kevin Winker, "ornithology itself became a scientific discipline."[63] Kumlien's collections in the 1840s and earlier, along with the bird collections of his contemporaries, were the resources on which the science was made. Bird collections contained data without which, in those early days, the scientific study of birds would have been unfeasible. The collection of bird specimens made possible the close observation, description, comparison, and classification that is the basis of ornithology. Kumlien likely took a small telescope with him, but if he or any naturalist wanted to know what birds looked like, how they were marked, the birds had to be brought close, and the only way to do that, generally, was to net them or shoot them. It's also important to remember that, in the 1840s, no bird population was threatened by the small number of naturalists shooting them for study. The birds Kumlien and Cornell killed would go into Löwenhielm's personal collection for study, or they would be sold by Löwenhielm to a school collection for study.

Well preserved and well cared for, specimens like the ones collected by Kumlien and Cornell can last for hundreds of years. "Bird collections need to be viewed as a highly versatile and indispensable resource integral to the continued successful (and economical) pursuit of a wide range of subjects," Winker argues. "Biological collections establish an object legacy—continuing sources of data that are repeatedly tapped to provide answers to questions about birds and environmental conditions. Many of these questions were not even imagined by those who have built these collections."[64] Museum bird collections, which can date back to the early 1800s (bird skins have been effectively preserved for only about two hundred years), are being studied today by those who research the effects of industrial pollution, climate change, diseases, and genetics. The specimens that men such as Kumlien gathered in the 1840s are even more valuable today than they were then—something these naturalists at least partially understood but probably not to the extent that we do now. Generations of ornithologists have used what Kumlien and Cornell saw alive on Gotland Island in 1842.

After the Gotland trip, Cornell went back to the university at Uppsala, eventually receiving a medical candidate degree in 1849 and a medical licentiate degree in 1856. In the midst of his studies, he returned to Gotland to work as a quarantine doctor from 1850 to 1851. He then became the assistant physician on the frigate *Eugenie* on a round-the-world trip from 1851 to 1853. The purpose of this trip, commissioned by the Swedish government and the Royal Swedish Academy of Sciences, was to promote Swedish trade as well as to make nautical, astronomical, zoological, and botanical observations. The expedition's botanist was Nils Johan Andersson, who had been Löwenhielm's travel companion to northern Sweden in the summer of 1843. Cornell worked as a physician in many parts of Sweden. He was several times hired during cholera epidemics. For the last ten years of his life, he continued his medical career in western Sweden, where he died unmarried at age forty-nine.[65]

The year after the Gotland trip, on May 6, 1843, Löwenhielm received his postponed degree and was free to travel. Through his maternal uncle, Löwenhielm received an offer to travel with professors Johan August Wahlberg and Carl Henrik Boheman from Stockholm along with Andersson, then a student, on a summer-long botanical and entomological excursion to the Lule River Valley in northern Sweden.[66]

The two professors had gone ahead by boat. Löwenhielm and Andersson left Uppsala on the morning of May 18 in Löwenhielm's hunting coach. For some miles north of Uppsala, the two travelers were accompanied by Löwenhielm's half-brother Gösta and by their mutual friends Thure Kumlien, Gerhard Von Yhlen, and nine others. At Tunaberg, the friends and family said good bye to Löwenhielm and Andersson, waving their student caps and handkerchiefs until they could no longer see one another.[67]

4

LOVERS AND FRIENDS ON THE *SVEA*

1842–1843

The ties that bound Thure Kumlien to his homeland were many braided threads—a love of family, a connection to the lakes and woods of his childhood, strong ties to good schools and good friends. In 1842 and 1843, those ties were not so much loosened as they were replaced by dreams of a future in North America. Kumlien was powerfully attracted to America and to the life one could make there as a settler and naturalist. In 1843, the path to such a life appeared before him.

As Kumlien and Måns Cornell followed in the footsteps of Carl Linnaeus on windswept Gotland Island, they must have talked now and then about Gustaf Unonius's letters from Wisconsin. Having read many of the same books as Unonius, sung the same Romantic songs, perhaps even shared some of the same doubts about his future, Kumlien may have dreamed the same dreams of liberty, equality, and opportunity in America that Unonius had dreamed.

But on his way back to Uppsala from the Baltic islands, Kumlien's life was forever changed. This is the tale as Kumlien's granddaughter Angie Kumlien Main remembered it being passed down in the family:

> After Thure returned from this collecting tour, he visited a school-
> mate, who also belonged to the aristocracy. The parents of his friend
> lived in their summer home on a beautiful lake. As soon as family and
> guest were seated at the dinner table, the maids brought in the food.

The young girl, Christine, who waited on the guest was beautiful and modest in demeanor. Thure fell in love with her at once and could hardly keep track of the conversation at the table. All through his visit he tried to see her, but he was always repulsed. Christine, who was a daughter of a low-ranking officer in the Swedish army in charge of training cavalry horses, knew full well that she had no business receiving attentions from the guest of a family where she served as a maid. The mistress of the house soon saw how things were and sent word to Thure's proud family.[1]

We do not know the name of the friend whose house Kumlien was visiting, where he lived, the date of this meeting, how the lovers continued to meet, what they said to each other, when they agreed to marry, or who in the family objected to their affair. We know nothing that contradicts this story, but we also have very few details to corroborate it.

Who was this "beautiful and modest" maid? In Sweden, she was called Greta Stina Wallberg. In Wisconsin, she became Christine, sometimes Christina, Kumlien. But her elder sister, Sophia Wallberg, who came to North America with Christine, kept the name Sophia Wallberg for all of her long life.

Sophia and Greta Stina were daughters of Johan Wallberg and Brita Cajsa Andersdottir. Sophia was their oldest daughter, born October 9, 1809, two years after her brother Johan Eric. Carl, Ulrica, and Lovisa came after Sophia. On April 21, 1820, Greta Stina was born—the sixth child of Johan and Brita Cajsa. When Greta Stina, later Christine, was baptized at two days old, the witnesses were two farmhands and two maids who lived nearby. Between 1822 and 1837, after Greta Stina, four more children were born to Johan and Brita Cajsa—Anders Jacob, Hedvig Catarina, Per Ulric, and Matilda.[2]

Johan and Brita Cajsa lived in Ryttarhof, Wallby Rote, Tibble Parish, Uppland. Johan was a dragoon, a soldier in a mounted cavalry unit. But he was not a dashing cavalry officer; he was a soldier and a farmer who trained cavalry horses recruited for the army by his local ward, or *rote*.[3] The ward provided the dragoons with a uniform, a small piece of land, a cow, chickens, pigs or sheep, hay, and seed. Wallberg likely spent weeks away from his family working for the farmers of the *rote* that supported him,

This modern image features a dragontorp, or dragoon's house, which was provided by the Swedish military to those who raised horses for the cavalry. Christine and Sophia Wallberg were raised in a small house like this one. PHOTO BY TUULA AUTIO, UPPLANDSMUSEET

attending the regiment's training camps several times a year. He did not own the land he farmed; it was his to use only as long as he was in the army. If he died or became disabled, the family would be forced to leave. If he was away at war, even for years, his family would have to do his work. The Wallbergs, like many Swedes of the time, lived a poor and precarious life full of hard physical labor.

From an early age, the Wallberg sisters would have been responsible for their younger siblings and for chores in the house and barns. Though adults carried out the important work of plowing, planting, and harvesting, bread had to be baked, meals cooked, chickens and geese fed, eggs gathered, the cow milked, yarn spun and woven, clothing washed and dried, gardens weeded, chickens and geese plucked, the cow fetched in the evening, wood and water brought into the house—all activities the children helped with or took on when they were big enough. Nearly all of the Wallbergs would have been working most of the daylight hours. The Wallberg sisters would have been taught to be useful and productive not long after they learned to walk.

When she was five and a half, according to church records, Greta Stina "didn't know the words well"; therefore, she "started at Holmqvist." Holmqvist was a shoemaker in a neighboring village who must have been running an unofficial kind of school. Greta Stina and her older sister Lovisa attended Holmqvist together for a time. Though the state Lutheran Church encouraged literacy so that each citizen could read the Bible, only about half of Swedes in 1800 could read, and most of those who could were wealthier people than the Wallbergs.[4] The only schools available to the Wallbergs would have been offered by the parish or by an individual in his home. Later records say Greta Stina read well and that she understood Luther's Little and Big Catechisms.

When Greta Stina was six years old, her seventeen-year-old sister, Sophia, and older brother Johan Eric left the family home. They moved together to the village of West Tibble to work for the family of a tenant farmer—Sophia to be a maid, Johan Eric to be a farmhand. After the age of fourteen or fifteen, most children of soldiers left home to work as maids or farmhands, as did most children of the poorer class. Young people without other means commonly worked in others' households until they married, inherited their family's farm, or bought out their siblings' shares in their family's farm. In September 1831, when Greta Stina was eleven years old and Sophia had been working as a maid for five years, their mother, Brita Cajsa, died of dysentery, leaving a husband and five young children still at home.[5]

Three years later, Greta Stina's father, Johan Wallberg, married Maria Andersdottir, a widow with three children from an earlier marriage. Johan and Maria had two more children, one of whom died in infancy. When she was fifteen, on October 23, 1835, Greta Stina moved to Sigtuna and began work as a maid in the parsonage of Johan Olof Sörling. The next year, she moved to live with the family of the widowed schoolmaster, teacher, parish clerk, and organist Johan Herman Zetterholm.[6]

Greta Stina's life as a maid was difficult not only because she had to perform heavy physical labor for very long hours and low pay but also because maids had no workers' rights and were vulnerable to sexual harassment, sexual abuse, and rape. If a maid refused to have sex with her employer, she would likely be raped or lose her job, and if she got pregnant, she would most likely lose her job as well. There were many orphanages in Stockholm for the babies of unmarried mothers.

As far as we can tell, none of the Wallberg children received more than a few years of haphazard schooling. Four of the Wallberg sisters worked as maids. By 1838, Greta Stina had followed her elder sister Sophia to Stockholm and found her way into work as a maid in the wealthy household where Kumlien stayed as a guest in the summer of 1842.

At Uppsala University, Kumlien had meanwhile become reacquainted with Gustaf Mellberg, a friend he had known at school in Skara. Kumlien and Mellberg attended school together and would also travel to America together. Mellberg would still be a friend, as well as a neighboring farmer, when Kumlien died in 1888.

Mellberg was born February 23, 1812, on a farm named Gullskog in Habo Parish in the Swedish county of Skaraborg.[7] In 1821, the family moved to Bränninge Inn—an inn with several farms attached where about seventy people lived and worked, among them servants, maids, and hired men. Whether or not he performed manual labor, Mellberg grew up observing the complex workings of a large farm and inn.

After four years of study at Skara Gymnasium, Mellberg elected to attend the Royal Academy in Lund, a preparatory school. This academy prepared him for the study of theology at the University of Lund.

After completing his university studies at Lund, Mellberg took a job as a teacher in Huskvarna for a year. Then, in 1840, he decided to continue his education at the University of Uppsala. As the only one of his brothers who wanted to go to school and was allowed to go to school, Mellberg must have had a powerful hunger for learning, a powerful academic ability, or both—he was either a student or, briefly, a teacher from the time he was about eight until he left for America at the age of thirty-one. These years of study left him in serious debt.

At Uppsala, Mellberg was back in Kumlien's life, and he stayed in Kumlien's life until the end. Though Mellberg's interests seem more classical and theological than Romantic, his friends included Kumlien, Gunnar Wennerberg, and Gustaf Unonius, as well as other Romantic-minded students at Uppsala. Unonius, who preceded Mellberg to Wisconsin, clearly influenced his decision to go to North America.

Mellberg and Kumlien were in some ways unlikely friends—Mellberg serious and severe, Kumlien more social and artistic in nature—but an element of mystery often enhances a deep and lasting friendship. Mellberg

Thure Kumlien and Gustaf Mellberg, pictured here in middle age, were friends for most of Kumlien's life. WHI IMAGE ID 69869

apparently did not share Kumlien's passion for the natural world, but these two eldest sons grew up near each other and attended some of the same schools, and both lost their fathers when they were young. They trusted and relied on each other, especially in later years. They appear to be the sort of friends who complement each other's strengths and weaknesses.

Emigrating from Sweden in the mid-1800s was an extraordinary, courageous act. Though later more than a million Swedes would leave for North America, when Kumlien and Mellberg left, only a handful had gone before them. The path was not well marked. It was a hero's path.

At one time, Kumlien might have imagined himself following in the footsteps of Elias Fries, a family man well ensconced in a professorship. That future would have held a certain appeal, and Kumlien seemed to be headed toward it in the first half of 1842 when he was almost finished with his degree.

But then several things happened within a few months: the letters from Unonius describing life in Wisconsin arrived in Sweden, Kumlien traveled to Gotland, and then he met a woman and fell in love. Importantly, he fell in love with a woman whom he could not easily stitch into the upper-class life of a professor or, say, the manager of an estate. Kumlien does not seem to have seriously considered that he, as the oldest son, would return home and manage the family estate.

Early in 1843, Kumlien was trying to complete his degree at the university and to figure out how he and Christine could make a future together within Sweden's rigid class system. In his mind, also, was the happy memory of the collecting trip to Gotland and Unonius's glowing descriptions of Wisconsin.

It was at this time that Kumlien, Mellberg, and the Wallberg sisters, Christine and Sophia, read in the newspaper of a rare event: in May, a ship would be sailing from Stockholm for New York. The following notice appeared in several newspapers: "On the brig *Svea*—Capt. J. E. Nissen— accommodations may immediately be secured for passengers to New York."[8] Captain Johannes Eric Nissen (1802–1874), a competent promoter, planner, and sailor, had noticed the winds of change in Sweden. He was the first to advertise sailings to New York primarily for passengers rather than the shipment of raw materials and manufactured goods. Captain Nissen's series of advertisements for passengers to sail on his newly renovated ship were, according to historian Axel Friman, "unrivalled at the time as to frequency, form and extent."[9] The previous year, only three ships had sailed for New York carrying passengers. Among these passengers were eight Swedes, including Pehr Dahlberg, who may have been the first Swede to follow Unonius to Pine Lake, Wisconsin.[10]

The *Svea* had crossed the Atlantic to New York in 1841.[11] Captain Nissen had made at least one earlier trip to New York, in 1842, bringing back to Gothenburg from America wheat and leather. When the *Svea* eventually left New York in mid-October 1843, after bringing immigrants, pig iron, and leeches to America, it carried "415 bales of cotton, 60 barrels of resin, 34 casks of tobacco, 25 barrels of flour, and dyestuffs" back to Sweden.[12]

Though Nissen had planned to leave Stockholm on May 1 and then Gothenburg—or Göteborg, as it appeared on Nissen's advertisement— several weeks later, the ship didn't leave Stockholm until the end of May. On May 29, the following article appeared in a Stockholm newspaper:

The brig *Svea* under command of Capt. Nissen today at nine o'clock a.m. sailed from here with 19 persons of different occupations who are emigrants to the USA. The vessel will stop at Göteborg where a large number of emigrants—reported to number 40 or 50

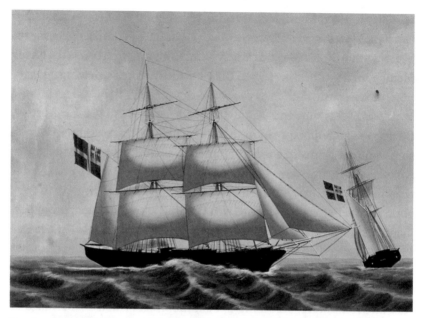

Kumlien, Gustaf Mellberg, and sisters Christine and Sophia Wallberg crossed the Atlantic Ocean on a brig named *Svea*, much like this one painted in 1850. COURTESY OF SJÖHISTORISKA MUSEET

persons—has assembled to participate in the journey. Also from Denmark will come 6 or 7 additional passengers. Having been renovated especially to accommodate passengers the vessel seems to be quite comfortable, and its commander is known to be exceptionally well qualified. The crossing is expected to take two months. Ticket prices are: 140 *Rks.* first class, 120 second, and 106 *Rks.* on third class. A doctor accompanies.[13]

Twenty-one-year-old Johan Olaf Liedberg acted quickly and purchased a ticket. His apprenticeship in Jönköping as a gold plater had just ended. He had promised his master—when he was twelve years old—that he would stay with him nine years in return for board, a pair of half soles, and one hundred crowns a year, "Nothing more."[14] Liedberg, who had survived a terrible time of cholera in his boyhood, had a prodigious memory for detail and wrote a lively memoir, which includes his trip on the *Svea* that also carried Kumlien and his friends from Gothenberg to New York.

Kumlien, Mellberg, and the Wallbergs would not be ready to leave when the *Svea* left Stockholm on May 29, but they would meet the ship in Göteborg by taking steamships south along the Baltic coast and then across Sweden through the lakes and the Göta Canal.

On May 15, Kumlien compiled an inventory of his possessions with the price he was hoping to get for each item. Trying to raise money for the tickets to North America, he listed for sale a chair and a cupboard, a book case, as well as the laboratory equipment he had used as a student at Uppsala University. He also listed a mounted swan, a partridge, an eagle, an eider duck, and a lark.[15]

Kumlien borrowed 1,333 Swedish riksdalers—a little more than five hundred American dollars today—from his friend Carl Gustaf Löwenhielm, who had paid for the Gotland trip. (Sixteen years later, the promissory note Kumlien signed on May 29, 1843, would be returned to him by Löwenhielm marked "Paid." Kumlien paid this debt over the course of these sixteen years by sending his friend collections of birds from Wisconsin.)

On May 31, in Stockholm, Kumlien collected three passports—number 1397 for "maiden Christine Wallberg," number 1398 for "maiden Sophia Wallberg," and 1399 for "student Thure Kumlien."[16] Mellberg had collected his passport on May 29. Kumlien and the Wallbergs stayed at an inn in Stockholm called Riga on Österlånggatan near where Mellberg was staying in Bollhusgränd.[17] From the funds he had borrowed from Löwenhielm, Kumlien loaned Mellberg the money for his passage.[18]

In the early hours of June 1, the four emigrants stood on the quay to board the paddle-wheeled steamer *Daniel Thunberg*. At five o'clock in the morning, they left Stockholm quietly, apparently with no fanfare, no procession of singing, cap-waving students, steaming out through the lakes, along the coast and the Södermanland archipelago.

Toward evening, they arrived in the bay known as Slätbaken; ate an evening meal of tea, beer, and some cold cuts; then arrived at the first lock of the Göta Canal late in the evening. With the boat tied up for a few hours, they retired to their cabins for the night.[19] While they slept, the ship went through the locks at Carlsborg. The next morning, they crossed Lake Roxen, passed through the seven Carl Johan locks, ate a midday meal at Husbyfjöl, and arrived at Motala in the evening.[20] After a calm night crossing of vast Lake Vättern, they passed through the locks at Forsvik, "steaming into Lake Viken."[21] Passing through their native province of Västergötland, both Kumlien and Mellberg must have thought of the younger brothers and sisters they were leaving behind, family they would probably never see again. Passing through more locks, they came into great Lake Vänern, the crossing of which took all night and part of the next day. They arrived at the first lock at Brinkebergskulle late in the afternoon and passed through the many locks at Akersberg in the evening.[22] The morning of their fifth day on the canal boat, they passed through more locks and finally made it to the Göta River. Kumlien, Mellberg, and Christine and Sophia Wallberg arrived in Gothenberg on the evening of June 5 and checked in to the Hotel Blom.

Liedberg, their soon-to-be shipmate, had arrived with a friend in Gothenberg on May 1, the advertised departure date of the *Svea*. "We stood in Gothenberg, paid for our tickets and were ready to go," he wrote in his memoir. "The captain, however, did not arrive before June 1st. During this waiting period I got a job in a mirror factory, so the delay cost me nothing, but instead I made a little extra money."[23]

The *Svea* finally arrived at the quay in Gothenberg on June 6. The next day, the Kumlien group at the Hotel Blom was informed that it would take a week to ready the *Svea* for the trip to New York. While they waited, they came to realize that others in their hotel were also sailing to North America on the *Svea*: the Reuterskiöld family (who changed their name to Reuterskiold in America) from Hössna in Västergötland, the Wahlin brothers, who were both blacksmiths, and Johan Olof Liedberg, the gold plater from Jönköping.[24]

On June 14, notice came to the passengers who had been waiting in Gothenburg for at least a week that they could now board the *Svea*, which was anchored out at Masthugg Quay. Liedberg wrote that it didn't take him long to get there, as he was eager to go. On board, he was "directed to the hold of the ship, a place that didn't exactly look like 100 crowns worth,—and without food!"[25] In the hold, Liedberg said, he was with a few other young craftsmen—"all skilled tradesmen, now thrown together with barrels and trash."[26] In first and second class, Liedberg noted, were "finer folk traveling with their families."[27] Kumlien, Mellberg, the Wallbergs, and the nine Reuterskiölds must have been among these "finer folk" settling into first- or second-class accommodations with their trunks and bags and food for the journey.

The brig *Svea*—belonging to shipowner C. G. Lindberg, who at that time owned ten ships—was a two-masted ship, ninety-eight feet long from stem to stern, and about twenty-six feet wide with two decks the length of the hull. The lower deck, or hold, ten feet high, carried Swedish iron and casks and barrels—and the third-class young men such as Liedberg. In the six-foot-high space between the hold and top deck was the forty-by-twenty-two-foot middle deck used as quarters for first- and second-class passengers.[28] Later, in 1846, when there was a much larger demand for passage from Sweden to New York, accommodations on similar vessels became more suited for passengers and likely much more crowded. In 1846, the brig *Charlotte*, about the size of the *Svea*, was outfitted with thirty sleeping bunks all around the hold, each seven feet wide, "each to accommodate 5 persons."[29] The *Charlotte*, in 1846, carried 150 passengers, followers of the charismatic Eric Jansson on their way to Bishop Hill, Illinois. The *Svea* carried about sixty immigrants.

In addition to the passengers, the *Svea* on this voyage was transporting about 275 tons of pig iron and, interestingly, a thousand leeches, probably the valuable and medically useful, though icky, European medicinal leech *Hirudo medicinalis*.[30] Pig iron, an important export from Sweden, cost about six dollars a ton to ship and was sold for between eighty and a hundred dollars a ton.[31]

On board the ship, still tied up at Masthugg Quay, as the captain waited for favorable winds, the passengers leaned on the rails, looking out to sea or back to shore at what they were leaving, wondering if they were doing the right thing.[32]

On Friday June 16, at nine o'clock in the morning, when the captain determined that everyone was on board, the sails unfurled and filled with a head wind that took the brig out to sea. "The mountains, the hills, the rugged coastline slipped away from us," recalled Liedberg, "and soon we could see nothing of our beloved homeland."[33] The last landmark "to which one may wave farewell" when sailing from Gothenberg was the cylindrical granite tower of the Vinga Island beacon.[34]

When the sight of their homeland slipped away, Liedberg noted, the group fell "into a deep gloom."[35] He said they realized that they were going to a place where they could not speak the language, and they wondered in what other ways they were unprepared. "Perhaps, too, we should never see our parents again," he wrote, "and our beloved Sweden was lost to us forever." Yet, he said, this general depression lifted "as we felt the winds of the open sea."[36]

The people on the *Svea* seemed to get along. "Oh, it could happen that a couple got into a fight over a card game," Liedberg wrote, "but the argument was quickly settled over a couple of bottles of sherry wine."[37] However, once on a name day (a celebration that involves the association of common given names with days of the year) someone drew a pistol "and that put an end to the sale of sherry wine by the captain."[38] Liedberg quietly took up the slack, though, by selling a tonic high in alcohol content called Hoffman's Drops to the men in first and second class who came down to the hold asking to buy rum. "This little business increased my purse," he said, and he got to know the men purchasing his tonic, "so I was later on able to go to Wisconsin with some friendly emigrants whom I met aboard ship and who planned to buy land in that state."[39]

According to Liedberg, the weather was good all the way west to New York. He noted just one incident, when there was a near collision at sea because the night lanterns were not up on the *Svea*. But Kumlien's granddaughter Angie recalled a different version of the voyage that had been passed down through the family: "For the first ten days all went well with the 'Swea.' Then they were overtaken by a violent storm, which the old ship could hardly weather. The passengers gave up all hope of ever seeing land again. In time, however, the storm abated, then the ship fell into a dead calm and drifted about for weeks without making any headway."[40] There may have been a storm that frightened the passengers, but the *Svea* certainly didn't drift for weeks without making any headway because it arrived in New York, if not in record time, exactly as planned. And the *Svea* was apparently a well-maintained ship, which made at least one more crossing of the Atlantic in 1844. Perhaps the story of the storm had simply been exaggerated in the retelling over the years.

"We had good music on board," wrote Liedberg, "and when the weather was still the young people danced on deck."[41] Kumlien may have played the flute he brought along, and passenger James Worm may have played one of his musical instruments. After eight days at sea and perhaps after a storm, on June 24, St. John's day, the whole ship celebrated Midsommar and the captain's name day, likely with music and dancing. Also, a laudatory poem was handed to Captain Nissen on his name day on behalf of the Danish and Swedish passengers.[42] Written in Swedish with an *ababcc* rhyme scheme by one or more of the classically educated passengers, the poem "certainly seems influenced by the romantic nationalism of the era."[43]

> Today a folk fest is celebrated in the Nordic lands!
> We emigrants are celebrating it too;
> It is a flower fest on our native soil—
> Though flowers cannot be picked on the blue waves,
> We have other reasons to celebrate this day:
> We are bent towards filling the need of the heart.

The poem continued, celebrating the "man who now is bringing us to the new world," the captain who would conduct them all to "peaceful harbor, / Where any kind of work will bring fruit." The poet wrote of the harsh fate of those who leave their native countries and friends "almost like outlawed men." Yet, until they reached the American shore, the ship was their country and the captain their friend. Comparing the captain to a Viking commanding a ship shaken by the Greek god of wind, the author of this fantasy attributed to the captain the ability to rule with wisdom "the wild winds amidst the whirl of the dance!" Ultimately, the captain was admired and honored by those passengers who were amused to recall their study of mythology.

They sailed day after day, week after week. Once in a while they saw another sail and occasionally a whale.[44] "As we sailed," Liedberg said, "we discussed our destinations and those who were going to Wisconsin decided to stay together."[45] Perhaps it was during one of those discussions that someone, maybe the captain, maybe Kumlien, brought out a map of North America. Because of the Unonius and Friman letters, we know many in the group were predisposed to go to Wisconsin. The story is told by Kumlien's granddaughter that on board the ship, the group looked at a map and Kumlien, seeing a large lake in southern Wisconsin on what might be a bird migration route, steered the group toward it. We do know that passenger James Benneworth, who spoke both Swedish and English, was *returning* to Lake Koshkonong. It's likely that both Benneworth's firsthand knowledge and Kumlien's interest in a lake like the Lake Hornborga of his childhood steered them to the shores of Lake Koshkonong.

In Kumlien's granddaughter Angie's account of the voyage, "Their food grew stale, their drinking water very low and almost unfit for use. When the drinking water finally gave out, their suffering was great. When they were fortunate enough to have a fog, they would hang out cloths to catch moisture from the air and then wring them to get a little precious water to alleviate their thirst."[46] They may have run low on water, as the business of transporting scores of passengers was new at the time, and the captain, inexperienced in this type of provisioning, may have underestimated the amount of water needed.

On August 16, a clear morning after eight weeks at sea, in exactly the time advertised, Liedberg remembered, "the captain called, 'Land to the West!' and there we saw a dark strip on the horizon. What joy this meant as the outline became more clear."[47]

Harbor pilots steered the *Svea* into New York Harbor. This was before the Statue of Liberty was dedicated in 1888 and Ellis Island opened in 1892 and even before Castle Garden opened in 1855 to process immigrants. Moored out in the harbor among a sea of masts, the immigrants aboard the *Svea* peered at the low crowded buildings of New York. They waited a day to be cleared by a doctor before they were put on a small sailboat, which took them to a dock, and then they set foot on the shores of North America.[48]

That first night, those traveling together to Wisconsin stayed in a small hotel. The next day, they took a steamboat up the Hudson River to Albany where, contrary to Kumlien's granddaughter's assumption, they traveled by railroad rather than the Erie Canal to Buffalo. Liedberg's detailed and believable account explained, "We went aboard a train to Buffalo in freight cars that were very unpleasant. The train was filled with many immigrants and the odor was indescribable."[49] Liedberg did say that the scenery was beautiful as they took this two-year-old train service roughly along the route of the Erie Canal. But the canal would have taken them four to eight days; the train, though it stank, was faster—it probably took just three to four days—and likely cheaper.[50]

Once in Buffalo, the Kumlien group made arrangements to take a steamship to Wisconsin, a trip that would begin in Lake Erie and take them up the length of Lake Huron, through the Straits of Mackinac, and down the west side of Lake Michigan directly to Milwaukee. But first, waiting for the steamship, they took a day to be tourists and visited Niagara Falls.

When they boarded the steamship in Buffalo after their visit to the falls, Liedberg wrote that the ship "was filled with immigrants from everywhere, people of various races and colors."[51] He said some people came on board who had been in Buffalo selling "pelts and leather goods." These people, with "their stately figures," wore "beads and bands of bright colors" as well as "clothing of skin and their faces were painted." The relatively uneducated Liedberg did not know what sort of people they were. He asked Kumlien, who told him that they were "Indians," and he was "puzzled no longer." When the steamship got to a point of Lake Erie "where the waters narrow to a channel the Indians left the boat." Liedberg and most others on the boat, no doubt, watched the Indians lower "small narrow birchbark canoes" into the water and load into them the bolts of cloth and other trade goods they had acquired in Buffalo. This occurred when the steamship was very near the Canadian border. "Through the trees," Liedberg wrote, "we could see a great many more Indians who had come to meet their friends among the travelers. They expressed their joy in dancing and shouting."[52] Finally, when the steamship began to move, he said, "we all stood in amazement gazing at the wonderful panorama on either side."[53]

Having traveled north the length of Lake Huron, the steamship carrying the *Svea* Swedes to Milwaukee stopped at Mackinac Island for one day in August. From the harbor below, the passengers looked up at Fort Mackinac where US Army troops were garrisoned. Liedberg recalled that two soldiers from the fort came on board the steamship. One of the soldiers asked, "Is there any passenger here who is a Swede?" "Yes," said Kumlien. When the soldier and Kumlien began to talk, they discovered that they had been students together when they were boys at Skara Gymnasium, an ocean and half a lifetime away. The soldier was Otto Vilhelm Olsson, the farmer's son who had been unfairly expelled from Skara in 1835. Olsson recounted how, when he arrived in New York, "he had been taken in and had lost all his money." According to Liedberg, Olsson told Kumlien that he did not have "the ability or the training for hard labor, and that he could not speak English."[54]

"We Swedes were deeply moved by this experience," wrote Liedberg.[55] They must have wondered if they, too, would be "taken in" and lose all their money. Would they have the ability for hard labor? Would they learn to speak English? But, Liedberg wrote, "At that moment the ship's bell signaled farewell." The lake was quiet, and they continued their journey without stopping until they reached Milwaukee.[56]

The Swedes from the *Svea* traveled down the western side of Lake Michigan. The shore they steamed along was part of the Wisconsin Territory, with its bluffs and bright beaches and clean bays where wild rice grew. They saw the limestone cliffs of the Door Peninsula, two or three settlements, and mile after mile of magnificent cedars and birch, pines and oaks and maples of the great unspoiled northern forest.

5

THE SWEDISH GENTLEMEN AT LAKE KOSHKONONG

1843

Of the sixty or so passengers on the *Svea* who crossed the Atlantic in the summer of 1843, nineteen of them continued on to Wisconsin by railroad and steamship. Of those who settled near one another and near Lake Koshkonong, Thure Kumlien, the Wallberg sisters, and Gustaf Mellberg were the only ones who had made plans to leave Sweden together. Sometimes later referred to as a colony or as a group of "Swedish gentlemen,"[1] the group had only just formed on board the ship. On the *Svea*, friendships were forged on the basis of similar education, background, and intention, melded by propinquity and the shipboard necessity to share water, food, and space.

The friends "who kept together during the hard pioneer days,"[2] as Angie Kumlien Main referred to them, were Kumlien; Christine and Sophia Wallberg; Mellberg; bachelors Sven Gabriel Björkander and Carl Gustaf Hammarquist, who went first to the Unonius settlement at Pine Lake but later rejoined his friends at Lake Koshkonong; a bilingual Englishman, James Benneworth, and his mother Alice Benneworth, who translated for the Swedes as they made their way across the country; and James Worm, an instrument maker from Denmark who later became a good friend of Kumlien's and helped him collect birds. Johan Olaf Liedberg, the young gold plater, stayed with the group for a year, then moved to Jefferson, Wisconsin, in 1844.[3]

Nine members of this party were the Reuterskiold family—Carl Edvard Abraham Reuterskiold; his wife, Maria; and seven children—who brought not only cash but a wagon packed with tools and household goods from Sweden. Born in 1796 in Västmanland, Carl Reuterskiold belonged to the nobility (but not the royal family, as was later inscribed on his tombstone). In the military until 1822, Reuterskiold then farmed in Hössna Parish. In 1821, he married the daughter of a priest. After they lost a month-old child in 1822, they divorced. Reuterskiold then moved to a farm in Björstorp in Hössna Parish where he lived with Maria Lindstrom, a *piga*, or maid, who was born in 1806. After having several children, they married and eventually had eight or perhaps nine children.[4]

That these passengers from the *Svea* were all going to the Wisconsin Territory seems to have been clear. But why didn't they go with Hammarquist and Bjorkander to Unonius's settlement at Pine Lake? Perhaps the name Pine Lake clued them in to the poor and dry soils found where pines grow. And the Swedes, who either knew or knew of Unonius, may not have been drawn to his strong personality. They likely decided to go to Koshkonong while they were on board the ship, having been influenced by Kumlien's interest in a lake such as Lake Hornborga and by Benneworth's previous knowledge and descriptions of the area.

In late August 1843, after no more than twelve days of travel from New York Harbor, this travel-weary group of Old World Europeans arrived in Milwaukee. They landed at the new pier at the foot of Huron Street,[5] the pier extending twelve hundred feet into Milwaukee Harbor. Built to bypass the sandbar between the harbor and the Milwaukee River, the pier accommodated the shipping of iron ore, lead, and lumber and welcomed dozens of German and Irish immigrants almost every day. As soon as they stepped ashore in this not-yet-ten-year-old boomtown, the immigrants were accosted by runners "thundering down the hollow plank walks,"[6] vying noisily to bring the immigrants up from the harbor to the hotels and boarding houses that lined Huron Street. Many Milwaukee citizens also swarmed down to greet the passengers and to collect and send packages and letters.[7] Two-wheeled horse-drawn baggage drays rattled down to the pier following the runners, lining up like today's taxi cabs. After the dazed newcomers sorted out their trunks and bags, and once their

possessions were piled high on a dray, they trudged—still on sea legs—
behind the driver who walked beside the dray, the men with guns slung
across their shoulders and the women and children carrying the lighter
bags and bundles. The immigrants were on their way to a hotel or boarding house and their first night in the Wisconsin Territory. A few months
before Kumlien's group arrived, the writer Margaret Fuller saw the "tor-
rent" of immigrants arriving in Milwaukee, often in their national cos-
tumes, "all travel-soiled and worn." After a night in "rude shantees," she
wrote, they "walk off into the country—the mothers carrying their in-
fants, the fathers leading the little children by the hand, seeking a home
where their hands may maintain them."[8]

To the Swedish gentlemen, the raw young city must have been exciting,
confusing, and perhaps repellant. These Swedes had come from the medieval cities of Uppsala and Stockholm where the centuries had burnished
the stone and brick of treasured old buildings, many of which are still in
use. Milwaukee was being hastily built out of green wood and the distinctive
cream-colored brick made of clay dug right in the city, kneaded by oxen
tromping in clay pits, and fired in kilns along the river. Perhaps the Swedes
reacted to Milwaukee as did a doctor visiting that year from the east:
"Such a miserable-looking place selected for a settlement I had never seen
before." Milwaukee was, he said, "sand-banks, frog-ponds, clay hills and
river marshes, with the unpretentious habitations of probably five thousand people scattered around in a desultory sort of way"—perfect for ducks
and geese and "wild beasts," but not for human habitation.[9] What repulsed
an eastern doctor, however, may have attracted a Swedish naturalist.

"One must accept things calmly," Unonius had sighed at his first impression of Milwaukee in 1841.[10] He and his brother-in-law, guns on their
shoulders, had taken a walk around the city. "To be armed with guns to
take an excursion within a city sounds perhaps a little strange," he explained to Swedes back home, but Milwaukee, like most American cities,
"had been laid out on a rather large scale and . . . most of it was still nothing
but an uninhabited wilderness; furthermore, a kind of watery wilderness,"
swimming with ducks. Unonius was offered city lots that were "puddles of
mire" for prices that he thought would equate to "literally throwing the
money into the water or, rather, the mire."[11]

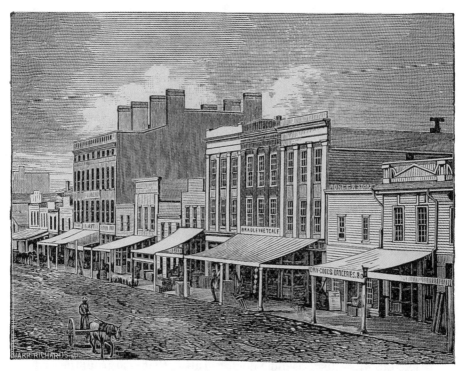

This woodcut, originally printed in James S. Buck's *Pioneer History of Milwaukee* in 1890, shows the west side of Milwaukee's East Water Street as it appeared in June 1844. YALE UNIVERSITY VIA HATHITRUST

Thure Kumlien the naturalist would have appreciated that Milwaukee was a fine spot for waterfowl, but Thure Kumlien the cash-strapped immigrant would have seen how expensive it all was. The lush and fruitful breadbasket that the bay and rivers and marshes of Milwaukee had been to Native people for thousands of years—a source of wild rice, waterfowl, fish, and shellfish—was lost on Americans and Europeans alike, as they transformed the site to a center of industry and commerce where almost everything one needed had to be bought and paid for with cash.

But the Romantic Thure Kumlien, who had time to walk around with Christine, more likely responded to "Milwaukie" as Fuller had responded: "This place is most beautifully situated. A little river, with romantic banks, passes up through the town. The bank of the lake is here a bold bluff, eighty feet in height. From its summit, you enjoyed a noble outlook on the lake." Exploring a narrow path along the lake shore where the ripples of the lake came up to their feet, Thure and Christine could look up to see the bluff as a "high wall of rich earth, garlanded on its crest with trees."[12] As did the curious and romantic Fuller, they might have climbed the path on the bluff to ascend into the lighthouse and watch the clouds over the lake and the steamships turning into the Milwaukee Pier.[13] Kumlien, who had collected plants since he was a boy, would have noticed the thickets of oak and wild roses "of so beautiful a red."[14] On this walk, Kumlien also would have seen something Fuller saw—"two of the oldest and most gnarled hemlocks that ever afforded study for a painter. They were the only ones we saw; they seemed the veterans of a former race."[15] The hemlocks were veterans of the nearly destroyed northern forest—the oaks, maples, yellow birch, white cedar, and white birch had been cut down to clear the city lots and to use for building and for fuel.

Though Kumlien and the other Swedes were passing through and heading to the country beyond, what they needed to set up homesteads would have to be purchased in Milwaukee. With even a cursory look, they could see that, rough as the town appeared, here they could get what they needed. Along the river and the one-mile industrial canal built by one of the three founders of the city, Byron Kilbourn, were flour and saw mills, tanneries and brick kilns. Below the sign of the big red boot, immigrants could buy their shoes and boots. Under the enormous teakettle, they could buy housewares. A silhouette of an anvil marked a blacksmith shop. Beneath the sign of the broad axe, they could get the edged tools necessary to clear their land. Leonard J. Farwell's hardware store on East Water Street sold to builders and householders under the sign of a mill saw and a stove. On the sidewalk in front of dealers in furs, immigrants stepped around a family of stuffed bears. The city had grown at an astonishing rate since it had been Solomon Juneau's fur trading post in the middle 1830s. There were carpenter shops and coopers and places to have wagons made, a tin shop and a paint store, clothing stores and tailor shops, furniture stores and print shops and jewelry stores. At this time, there were more "rum holes" than churches.

Though Kumlien couldn't know it at the time, there were men in that raw city who were dreaming of a museum in which to preserve the specimens of the natural world that were being so quickly destroyed. At about the time Kumlien arrived, Milwaukee botanist and all-around scientist

Increase Allen Lapham had requested money for a Milwaukee museum from Massachusetts botanist John Bartlett. Bartlett's reply was that "there are not people enough yet in your country to support anything of the kind. . . . A News Room might do in your place but not a museum."[16]

Milwaukee was not yet a big city, but to Kumlien and the other Swedish immigrants, it was a foreign city. Luckily, there was a countryman who would guide them through the strangeness—a Swede named Oscar Lange who worked in Leonard Farwell's hardware store. Lange was the man who had shown Unonius the ropes when he arrived. Upon hearing of another group of Swedish immigrants, Lange advised his countrymen in the purchase of land and probably also let them know what they would need to purchase to set up farmsteads out in the territory.[17] The very sociable Lange and his Irish immigrant bride almost certainly put themselves out for the travelers—countrymen and fellow immigrants.

After Kumlien and the Wallberg sisters found—perhaps with Lange's help—places to board in Milwaukee, Kumlien visited the seven-year-old white frame courthouse surrounded by grass and young trees and a picket fence. Here he "declared his intention of becoming an American citizen in the Milwaukee circuit court, territory of Wisconsin, on August 28, 1843,"[18] a first step for immigrants in buying government land.

On September 5, 1843, Thure Kumlien and Christine Wallberg were married by Justice of the Peace William A. Prentiss, a merchant and politician who would one day become the mayor of Milwaukee. The couple had known each other a little more than a year, but in that year, they had decided together not only to marry in spite of their class differences, but also to leave the country of their birth and cross the ocean together. While they knew little about what their future would be, they knew they loved each other. We can imagine that Christine and Sophia had played a quiet part on the voyage, where the talk would probably have been dominated by the educated and upper-class men. But once Christine was Thure's wife and they were in this new world where the life of the mind was second to the needs of the body, Christine's knowledge and skills would be very important. In September 1843, the Kumliens did not yet know where they would live or which aspects of their previous lives would be held onto and which would be left behind. Thure brought his classical education and his

At Milwaukee's first courthouse, built in 1836 by Solomon Juneau and Morgan Martin, Thure Kumlien declared his intention of becoming an American citizen on August 28, 1843. Eight days later, Thure and Christine Wallberg were married—either in the courthouse or in the attached office of the justice of the peace. PHOTO COURTESY OF THE MILWAUKEE COUNTY HISTORICAL SOCIETY

romantic dreams of the forest and of collecting birds. Christine must have brought dreams of a safe and dignified life as a wife in her own home rather than as a servant in someone else's home. There are no records of what these first days of their marriage were like, but their joys must have been tempered by trepidation.

Sometime in the middle of September, Thure left Christine to travel west with the other men of this group to select the land where they would settle. Christine, Sophia, and Maria Reuterskiold and her young children stayed behind in Milwaukee.

After they left Milwaukee, the Swedish gentlemen walked through beautiful country. On a wagon track through about ten miles of old and dense oak, beech, and maple forest, they passed scattered clearings and cabins. According to settler Edward Holton, "Little or no work was done on [the roads] beyond making sufficient track for wagons to wind along, and poles and logs were thrown across the streams and swamps. . . . Still the great army of immigrants . . . made their way through these tracts to the more open and genial country behind."[19]

The roads were bad through the woods, but Kumlien would have been looking at the trees as Unonius had two years earlier. "Nothing can be compared to an American autumnal forest," Unonius wrote. "All possible tints and shades, from bright red, gleaming crimson, scarlet purple, and orange, to dark green and dark brown, form a picture one can nowhere else behold."[20] Kumlien's granddaughter Angie Kumlien Main imagined him as a "young man of only twenty four with the soul of a poet and an artist, walking through the forests with their oak openings during the beautiful autumn days." Kumlien, she speculated, "must have gazed with rapture upon the brilliant colorings of the fall flowers and foliage. How he must have wanted to follow and identify the strange birds as they flew from tree to tree, but he had to keep up with the other men and follow the narrow wagon road toward their homes-to-be . . . in these beautiful woods near Lake Koshkonong."[21]

East of the Rock River and north of the Bark River, they passed through a forest of what Lapham called "some of the finest trees in the Territory."[22] The white oak trees would have been familiar to Kumlien. After they crossed the prairie near Whitewater and Scuppernong, the remainder of the country they crossed was "openings," or savanna—rolling parklike oak openings with "scattering trees of the kind of oak here called bur oak."[23] The ragged and emphatic bur oak, which Increase Lapham called the most ornamental of Wisconsin oaks, may have been new to Kumlien, who was used to the more formal and sedate white oak. Because the American Indians had controlled the underbrush by intentional burning in the savanna, the "ground everywhere was as clean as the prairies and the timber or oak trees that made up the woods were wide apart, presenting a scene more like an orchard than a timber land," settler Lucien Caswell wrote. He continued that in 1837, they "could drive all over the land and through the woods everywhere as easily as one could on a traveled road."[24]

This oak opening or savanna in southeastern Wisconsin was drawn by Adolph Hoeffler, a Romantic German traveler and painter, in 1852. WHI IMAGE ID 31186

After walking more than sixty miles and spending one or two nights in taverns, the Swedish gentlemen finally stood on the shore of Lake Koshkonong. Liedberg wrote, "We had reached the most beautiful place that I had seen in America."[25]

In this beautiful place, the Swedes soon met a small and congenial group of settlers who had been living there for a few years. Elias Downing, a friend of James Benneworth, had arrived in the early 1840s. The Devoe family had come from Michigan and settled near the Indian-Army trail, which is now Highway 106. The Samuel Kirby family from Connecticut homesteaded near the Devoes. As Main noted, the "Swedish pioneers were not without a few neighbors, and time proved them to be very good ones."[26]

Over the following weeks, the Swedish gentlemen selected their land by locating unclaimed land within blazed section markings, marking their own selections, and registering their claims in the land office in the county seat of Jefferson. Carl Reuterskiold bought and paid for 320 acres.[27] Kumlien selected forty acres on an Indian trail. This wooded forty in Section 18 of Sumner Township was located on a ridge and, he thought at the time, included a spring. It was just south of a group of Indian mounds, which rose like the bones of a spine along the ridge—one effigy mound, two linear mounds, and twenty-four conical mounds, forming "an almost unbroken line along the crest of a rather prominent ridge."[28] From the ridge, they had a long view south and east toward Lake Koshkonong and over the lowlands along Koshkonong Creek. "To the west there is a rolling upland," noted archaeologists surveying the area in the early 1900s. "To the east gentle slopes lead down to lower levels of the creek bottom."[29] Below this ridge was the sheltered south-facing spot Kumlien selected for his cabin site. Mellberg's less interesting and perhaps more productive eighty acres were a mile and a quarter east of Kumlien's.

Like Caswell and many other settlers, the Swedes passed up prairie land and chose forested land, believing that the trees were an indication that the soil would be rich for farming. "We admired the great prairie for its beauty and magnificent appearance," Caswell wrote, "but were skeptical as to its productive qualities. . . . We felt that land which did not grow timber certainly would not grow anything else."[30] "Besides," wrote Main of her grandfather and the other Swedes, "they liked trees and had been used to them at home."[31] They may have wanted to be near Koshkonong Creek, which would provide power for a sawmill. "Another thing that was attractive to them," Main added, "was the natural grass on the marshes which could be made into hay."[32]

Before they returned to Milwaukee for the women and children, most of the men in the party built two temporary log shacks to get them through the winter. Johan Liedberg, who stayed with the Swedish gentlemen for a short time before he moved up to Jefferson, wrote that they filled the cracks between the logs on the first shack with grasses and put grasses on the roof: "When finished it looked like a haystack."[33] But Kumlien made arrangements to stay in the Downing family's hunting shack on the shore of Lake Koshkonong in a grove of bur oaks.

Opposite: The highlighted points on this 1846 map by Increase Lapham show the places in Wisconsin Kumlien traveled in his lifetime, almost all in the southeast portion of the state. WHI IMAGE ID 92358; MODIFICATIONS BY MAPPING SPECIALISTS

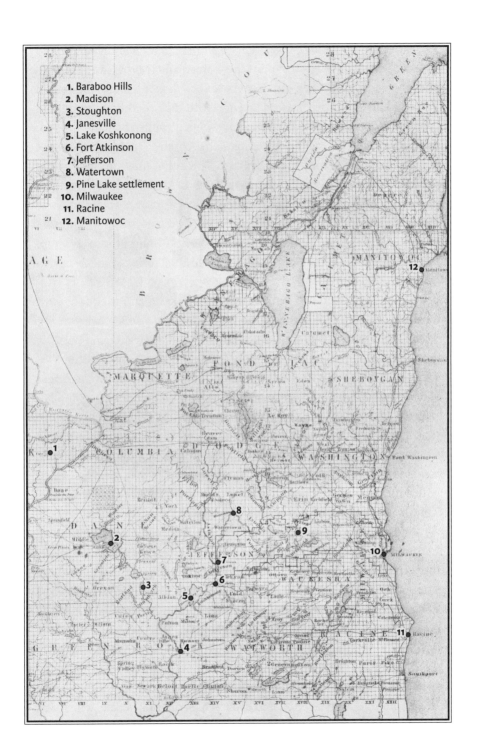

1. Baraboo Hills
2. Madison
3. Stoughton
4. Janesville
5. Lake Koshkonong
6. Fort Atkinson
7. Jefferson
8. Watertown
9. Pine Lake settlement
10. Milwaukee
11. Racine
12. Manitowoc

Sometime in October, the Swedish gentlemen walked the sixty or so miles back to Milwaukee to collect the women and children and the supplies they needed.[34] They purchased a wagon, loading it and the wagon Reuterskiold had brought from Sweden with tools, bedding, household supplies, flour, and pork. Both wagons would be pulled by pairs of oxen. Beds were made for the littlest Reuterskiold children in dry goods boxes in a wagon. Sadly, several of the children had caught smallpox. According to Main, "The first night out the party tried to get lodgings at a settler's house, but when one of the sick children looked up over the box and it was discovered that the travelers had smallpox, the whole group was refused shelter."[35] Later that night, farther along the road, one of those children, most likely one-year-old Conrad Reuterskiold, died and was quickly buried by the side of the road.

Back at their land in Koshkonong, Reuterskiold and Mellberg built dwellings to get them through the first—unusually mild—winter. But Kumlien, instead of building, spent the first fall observing and hunting the migratory birds on the shore of the lake while he, his wife, and his sister-in-law stayed in the hunting shack until the weather was too severe and they moved into the Downings' log cabin. Because the winter of 1843–1844 was mild with little snow, the young Swedes camping in little more than a lean-to must have had a fine time living out the romantic dream of a cottage in the forest. Liedberg, living near the Kumliens that winter, remembered that "we were not without food since the forest was filled with wild animals and deer, ducks, and fish which could be had with ease. Kumlin [sic] was a good hunter."[36]

Kumlien was on one of the great North American flyways, where bird migration was funneled southwest, initially along the Niagara Escarpment from Green Bay and its marshes to the marshes of the shallow Lake Winnebago to the vast Horicon Marsh and then to the lush Lake Koshkonong—a route following what is called the Rock River–Lake Winnebago–Green Bay Lowland.[37] Arriving at Koshkonong, whether you were a bird or a Swede, you might have thought you'd arrived in the south. In truth, you would have arrived at the northern edge of a country different even from Milwaukee. The trees were different from northern trees—more silver maples, American elm, and black ash. From Milwaukee, which was

Beyond these unidentified nineteenth-century waterfowl hunters and their dog, you can see a marsh like the Koshkonong marsh where Kumlien hunted. WHI IMAGE ID 59473

at the southern edge of the northern forest along Lake Michigan, the immigrants had come through oak savanna and open prairie to a lush place that was, though still the Wisconsin Territory, climatically at the northern edge of the Illinois prairie.

Though Kumlien did clear a bit of land for a field and collect some logs for a cabin, he spent most of that first fall as a scientist and collector rather than a farmer. Main wrote,

When Lake Koshkonong was finally reached, . . . it must have seemed as though paradise was spread out before him. The primeval forests all about him were alive with songbirds; the mudflats on the lake shore were covered with sandpipers at this time of the year; rails worked their way through the reeds; blue herons fished on the shores as they do now; ducks of many kinds covered the lake; whistling swan sailed majestically on the water; wild geese honked their way overhead; bitterns were almost invisible in their old-stump camouflage; red-winged and yellow-headed blackbirds teetered on the reeds.[38]

That fall, Kumlien witnessed for the first time one of the great migrations of the north-nesting birds to their southern feeding grounds—Canada geese; the voluble sandhill cranes, so similar to the common cranes of his childhood; the mallards and blue-winged teals and wood ducks—so many birds, skimming over the open water at the center of the marsh. Borrowing a canoe to go out on the lake, he pushed through the bending stems of the wild rice. Though we can't see it today because the Rock River was dammed to raise the level of the lake, Kumlien had landed himself in one of the great wild rice marshes of North America, a breadbasket for Native people—as well as birds and animals—for thousands of years.[39]

Though most American Indians had been forcibly removed from this land of theirs during and after the 1832 Black Hawk War and the removals of 1841, American Indian people still lived in the area when Kumlien arrived, and they harvested wild rice in canoes on the marsh. A few years later, Kumlien would have witnessed one of the last—though diminished—wild rice harvests before the Rock River at the foot of the lake was dammed and the higher water level slowly killed off the rice, changing forever the character of the lake. The plant we call wild rice, technically a grass rather than a rice, was named by Carl Linnaeus. The Linnaeus apostle Pehr Kalm had brought it from North America to Sweden in 1753. Linnaeus named it *Zizania aquatica*—*Zizania* for wild rice and *aquatica* because it grew in the water. In Kumlien's botanical studies and his reading about North America, he likely would have learned of it. Growing three to five feet above the water surface, native wild rice plants could produce one hundred fifty to two hundred pounds of rice per acre. The fifteen square mile, ten-thousand-acre lake would have yielded over a million pounds of rice in a good year, giving the people who harvested and stored the rice potential food security. And below the surface of the lake grew the grassy wild celery plant, the buds and blooms of which were particularly sought after by diving ducks.

It was the vast numbers of ducks—canvasbacks and redheads and many others—that most impressed Caswell and the Swedish settlers in those early days. When the ducks came to Lake Koshkonong by the millions to feed, in Caswell's words,

> they would light down all over the vast rice fields and feed on the unlimited quantity till they were fat and most delicious food. It was only a question of ammunition in the number one would kill at a shooting. When a gun was fired there followed every time a sight to behold. The noise of the gun would stir them up and they would rise out of this field of rice in such quantities that the roar was like distant thunder. The atmosphere overhead would be filled till the sun at times would be almost darkened. No tongue or pen can describe the number or quantity.[40]

In a bur oak grove on the shore of Lake Koshkonong on a September day in 1843, Kumlien must have stood looking out over a lake that contained more water fowl than he had seen or heard at one time since he had left the Swedish Lake Hornborga of his childhood. Yet in Sweden, he knew the names of the birds he heard and saw. How would he learn the names of these American birds? He had no book of American birds.

Kumlien would have heard a quiet, unfamiliar chittering in the tops of the oaks—a flock of little brown birds making its way almost unseen through the treetops, busy at their hunt for insects, busy at making their way south to their wintering grounds. He might have wondered not whether he would see them again, because he knew they would be back in the spring, but whether, by then, he would know what they were.

And then he could have heard a familiar cry—the rare black tern that he and Måns Cornell had finally seen on Gotland Island. Here they were: hundreds of them, migrating, hovering over waves as terns do, but in this case over waves in the sea of wild rice. He would have watched them dip down now and then, swallowlike, to snatch an insect from the slender, swaying tops.[41] Though he had come all this way to see new birds, birds unknown in Sweden, Kumlien must have been pleased to see and hear a familiar bird.

Despite the strangeness of this place, the scientific and poetic mind of Kumlien must have known that here was home. Everything conspired to bring him here—to this Eden of bird life: his boyhood interest in natural history, his reading, his studies at Uppsala University, his collecting on Gotland. Also, his friends Unonius, Löwenhielm, and Mellberg and his love for Christine Wallberg. Kumlien must have recognized that here was, for now, an unspoiled land where bird life and plant life existed as they had for thousands of years, where no one had drained or dammed the waters, where above the lush and delicate plants of the bog and the marsh and the upland, the great old trees still stood.

6

A Settler's Journal

1844

The rugged individual single-handedly taming the wilderness was likely the romantic image held by Thure Kumlien and the other Swedes before they left their home country in the 1840s. Kumlien may have imagined that he would go into the forest and mark his claim in the bark of the great trees. He and his wife would build a log cabin. They would be the first people in that wild place. Everything they needed they would have to make or grow or find in the woods. Only occasionally would their isolation be relieved by visits of other white settlers. But what they found when they arrived at Koshkonong was very different from this common romantic vision of settlement.

Kumlien and the other Swedes from the *Svea* had not arrived in an empty quarter. Though they made claims on land the US government called "unoccupied," the area around Koshkonong had been hunted, fished, and planted for at least a thousand years by Native people. And much of the land surrounding their claims had recently been settled by other immigrants—New England Yankees and Norwegians. The Swedes likely found the place to be more complicated than they expected—richer in its long and recent history of occupation by American Indians and its current occupation by other settlers. As Kumlien found the bird life here richer and perhaps more overwhelmingly abundant, he must also have found the settler's life to be more challenging than he expected.

In the winter and spring of 1844, Kumlien cleared land, split fence rails, and followed a borrowed plow and oxen. He was twenty-four years old, and there were difficulties on every hand—his inexperience with heavy labor, extremes of weather, and no cash. His life in Sweden had prepared him to read and think and talk and play music, to be a friend and a naturalist, but it had not prepared him to be a farmer. Kumlien was not alone.

From the mild autumn of 1843 to early 1844, Kumlien and his wife, Christine, and Christine's sister Sophia Wallberg lived in a hunting shanty belonging to Elias Downing on the shore of Lake Koshkonong, steps away from the more substantial log cabin where Downing lived with his wife, Polly, and their three children.[1] The crude hunting cabin in an oak grove where the Kumliens camped had a stone fireplace, which might have made it comfortable in mild temperatures. But during the coldest part of the winter, they moved into the warmer but crowded cabin with their new friends, the Downings. In the early spring, they moved back into the Downings' shanty.

By February 1844, Wallberg had begun living with other families who hired her to help in the house, nurse the sick, care for children, weave and knit, and work in the gardens and fields. The Kumliens had moved onto their land and were living in their own crude and temporary shanty about a mile northwest of the Downings' Carcajou Point cabin. A trail still led from the Ho-Chunk village site on Carcajou Point through Kumlien's forty acres on the brow of a ridge and past the effigy mounds just to the north.

On February 14, 1844, Kumlien began what he titled a "Work Journal . . . With Notices of Various Occurrences," which he kept, with some interruptions, for about six years. Writing in ink and pencil in a large ledger book, likely by candlelight on a table he would have made or on a lap board he would have split, Kumlien began writing—in Swedish—almost every day. He began his journal to keep track of the complicated exchange of labor, tools, oxen, and goods among his neighbors and friends. Kumlien noted work done on his own place and for the neighbors, and he noted birds and animals shot and preserved, fish caught, visits with friends, and sometimes, though not always, important family events. In 1844, Kumlien wrote one line a day, sometimes two or three or occasionally more, but he wrote enough about his days that his life in those early years begins to come into view. Though there is little musing in the early entries of the journal, his notes still convey what he and other settlers were doing and what that life was like as they made homes and farms in Koshkonong country. The picture that emerges is far from the stereotype of the pioneer family alone in the woods. The work was hard, but it was often shared. The Kumliens saw friends and neighbors almost every day.

Here is the first month of Kumlien's journal as it appears in Albert Barton's translation and typescript:[2]

W. Feb. 14—Chopped wood. Cut down five small trees for posts where the field is to be.

Th. 15—Do.[3] Made an axe handle (received clamp from Janson). Worked around shanty. Made a sawbuck and some clapboards.

F. 16—Do. Worked around shanty. Started making knives, forks and spoons.

S. 17—Downing here (Saturday). Hunted (two hours) forenoon. Shot 10 squirrels. Bought five bushels good wheat.

S. 18—To Ft. Atkinson (tea, tobacco and sugar, etc. for 95 cents).

M. 19—The white and old brown hen laid eggs. Fixed door and stairway. (Mellberg wrote to Bjorkander.)

T. 20—Thrashed at Downing's and Charrick's by Kyle's lake.

W. 21—Visiting.

Th. 22—20 foot rails, round and split. Sharpened axe and knife. One egg laid.

F. 23—10 foot rails. Do. (Stuffed a blue jay.) Charrick came here with new ale.

S. 24—10 foot rails. Do. Helped at bridge building till noon. Paid Janson 10 and Reuterskiold 40 lbs. fine flour.

S. 25—Shot a squirrel of a variety new to me. Tea at Charrick's. Light hen laid. Reuterskiold wrote to Sweden.

M. 26—Helped Charrick all afternoon. Janson saw six ducks.

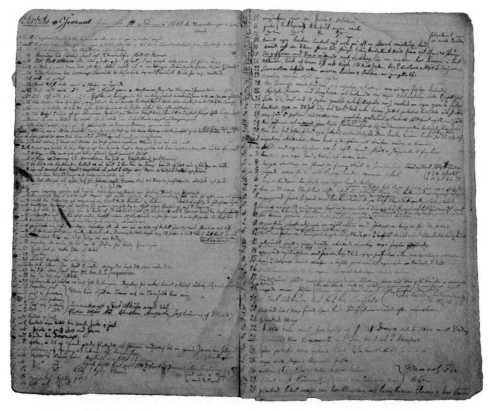

Two pages from Kumlien's Swedish-language work journal. THURE L. KUMLIEN PAPERS,
WIS MSS MQ, BOX 2

T. 27—Chopped five lengths of ash in the commons. Shot three
small squirrels. Fixed cow stable with a real stall. Both hens laid.

W. 28—Chopped a little. Chirsten [*sic*] arranged with Downings
for the pigs.

Th. 29—Saw two wild geese. Built at a hen house. Fia at
Downings.

Though apparently not a severe winter, still it was February in Wisconsin, so the days were short and gray, cold and damp. Thure and Christine
were living in a leaky shanty, but Kumlien must not have spent much time
in it. Almost everything he did was out in the fields he was clearing, in the
marsh, or in the woods.

Christine, on the other hand, had work to do closer to and in the home. She hauled water up from the spring and tended the open fire in front of the shanty where she cooked the black bass, redhorse, and sunfish Kumlien caught in the lake as well as the squirrels and geese and ducks and prairie-chickens he spied with the telescope and then shot with the shotgun he had brought from Sweden. They slept on a platform he made, on ticking she likely brought from Sweden, which she stuffed with marsh hay Kumlien cut with his scythe. Christine tended the hens they bought or traded for—the white one and the old brown one—and gathered, one by one, the eggs she and Kumlien delighted in. She milked the cow and churned butter, though sometimes Kumlien churned, too. Now and then, they walked the ten or so miles to Fort Atkinson and bought sugar, tobacco, or flat bread. Christine was young and strong and used to hard physical work, but she was also expecting their first child, so living in a rough dwelling was perhaps more of a worry to her than it was to Kumlien, who would rather have been in the swamp shooting birds anyway.

Kumlien's early journal makes clear the enormous amount of hard physical work he and other settlers engaged in—felling trees; chopping, hewing, and splitting wood; building fences and dwellings; working the soil and planting and harvesting; feeding, chasing after, and butchering chickens, cows, pigs, and oxen; fishing and hunting.

Almost every day Kumlien would have to cut a "daily portion of wood"[4] for heat and for cooking. Chopping wood with an axe, splitting it with a maul and wedges, then stacking it would have been daily chores, as was carrying water, too common after a time to even list in a journal, so after a short time, Kumlien didn't mention maintaining the essential wood pile. While clearing a field to prepare for planting, he was also felling trees that would be the log walls of the more permanent cabin he would one day build. On March 9, in what he called fine weather, he cut fifty-eight timbers from trees he had felled the day before—timbers he would use in repairing the shanty or in constructing the planned cabin. For a man who had been a university student less than a year before, learning to efficiently use a felling axe or a broad axe without slicing his leg or cutting off toes would have taken some time and earned Kumlien many good blisters. Axe handles sometimes broke, but Kumlien learned to make them from ash, using a clamp or "dog" he borrowed from a Norwegian immigrant neighbor. On March 22, he reported that he cut "20 foot rails, round and split."[5] And, probably using a froe that he struck with a maul, he split roofing clapboards. In those days before barbed wire was invented, splitting rails for building and repairing fences was constant work. Farm animals had to be penned so they didn't wander off, while garden plots and planted fields had to be fenced to keep out both domestic and wild marauders.

Most work with wood was heavy work, but Kumlien also carved—probably from ash—knives, forks, and spoons, though he still had a few pieces of fine old silver he had brought from Sweden. Kumlien also worked at building and improving the shanty, fixing a door, stairs, and a ladder using local woods and precious nails he'd bought from a blacksmith. And he worked on the cow stable and the henhouse, repairing doors and gates. The work on a neighbor's bridge across Koshkonong Creek or on communal bridges was wood work as well. Neighbors would gather at a creek at a designated time, bringing their own tools. Kumlien had an axe and a carving knife, but he sometimes had to borrow a hatchet from Reuterskiold. All of this wood work meant that Kumlien, and most settlers, had to spend a fair amount of time sharpening edged tools—which he mentioned several times in the journal in 1844.

Much of this work was ongoing, not tasks that could be completed once and forgotten about for the rest of the year or years to come. Even when a cabin was completed, the nature of the shallow and rough stone foundations and the proximity of the wood to damp earth meant that the dwelling, stable, henhouse, and fences were always in need of repair. And permeable. Mice and snakes found their way into the shanty and the cabin in the early days.

Considering Kumlien's interests and fine motor skills as he preserved birds, played the flute, and occasionally drew and painted, it's likely that he was better at carving spoons than splitting rails. And after his neighbors—and he—knew more about his abilities, Kumlien began to be hired now and then to build cupboards and shelves. In July 1844, Kumlien built a cupboard at Henry Charrick's with shelves for milk. A few years later, Kumlien took work in Jefferson in a furniture turning factory. He learned to use a wood lathe and apparently liked it, making many decorative pieces of furniture for their house.

In the 1844 journal, which covers only about seven months, the names of more than twenty neighbors are mentioned. These are mostly men—men who shared field labor and teams of oxen, men who borrowed, loaned, sold, gave, or hauled fence rails, flour, seed, pork, tools, ale, potatoes, and more. Kumlien and his neighbors had to trade their work, their produce, and their skills in order to survive. Theirs was a barter economy, not yet a cash economy. Not only had Kumlien not yet planted a cash crop—winter wheat—but at that time in Koshkonong country, hauling wheat using oxen and a wagon along the rutted roads to the nearest market was impractical and expensive. A farmer would have to borrow or rent a wagon and a team of horses or oxen, and he would have to pay tolls on some of the roads to markets in Whitewater and Milwaukee. The earliest white settlers on the south shore of Lake Koshkonong had selected their home sites to take advantage of planned Rock River steamboat transportation. But this water transportation on the lake and the Rock River had not materialized. And Kilbourn's and Lapham's Milwaukee and Rock River Canal, which would have ended at the confluence of the Bark and Rock Rivers at nearby Fort Atkinson, had been canceled as the more lucrative railroads took priority.

As might be expected, Kumlien's closest work and social connections were with Swedes—Gustaf Mellberg, Reuterskiold, and later, the Dane, James Worm.

Mellberg, still a bachelor at that time, and Kumlien lived a little more than a mile apart, and they shared the planting of corn and potatoes, hoed corn in each other's fields, and together prepared clean ground for threshing. On August 17, Kumlien drove a team of borrowed oxen to Mellberg's and finished Mellberg's plowing. On August 23, when Kumlien was sick, Mellberg helped him cut hay all day. But about a third of the entries that mentioned Mellberg in Kumlien's 1844 journal concerned matters other than the exchange of work. In February, Kumlien noted that Mellberg wrote a letter to their friend Sven Bjorkander.[6] There were visits at each other's houses on Sundays and, at least once, Mellberg came to dinner. On Sunday, August 11, Kumlien wrote that he was "at Mellberg's awhile and got some peas."

Though Maria Reuterskiold—Kumlien called her "Mrs. Reuterskiold" in the journal—came to visit Christine, there does not seem to have been any "visiting" back and forth between Reuterskiold and Kumlien, as there was between Kumlien and Mellberg. Kumlien's relationship with his older neighbor Reuterskiold seems to have been a more formal one. From Reuterskiold, Kumlien twice borrowed "fine meal" and a hatchet. Reuterskiold, who owned several wagons, a team of horses, and a team of oxen, hauled three loads of timber for Kumlien and, another time, one hundred stakes. On one day in May, Kumlien couldn't break ground because he needed Reuterskiold's oxen and Reuterskiold was off hunting them up. Twice Kumlien noted what he owed Reuterskiold, and once that sum was sixteen cents. But on Saturday, August 24, Kumlien reported that he was at Reuterskiold's cabin from eleven o'clock at night until five in the morning. Though he didn't say why, it was likely to sit with one of the family who was ill.

Kumlien met Worm, an instrument maker from Denmark, on board the *Svea*. When they had all first arrived in America, Worm had, instead of settling at Koshkonong, gone to Gustaf Unonius's settlement at Pine Lake. On March 10, Kumlien recorded that he wrote to Worm, perhaps to ask him to visit because later that month Kumlien noted that Worm was at their place. Then Worm must have moved to Jefferson, the county seat, because at the end of April, Kumlien stayed overnight in Jefferson with Worm. And Worm, with his wife, visited the Kumliens in May and July. Worm seems to have been more of a friend sharing an interest in music and the natural world than a settler Kumlien shared hard work with. Though they didn't live in the same township, Kumlien went out of his way to see Worm. And it wasn't long before the music-loving Dane decided to settle at Koshkonong.

Throughout Kumlien's life, he maintained relationships with at least a few friends, not all Scandinavians, who shared his intellectual, musical, and natural history interests. Kumlien also developed relationships with some of the Yankee and Norwegian settlers whose settlements he was surrounded by. He had good friends not just because good people happened to be his neighbors but because he was an attractive and attracting man, a man with whom people enjoyed talking and singing and laughing.

Yankees from New England and some English and Irish immigrants were the first white settlers in the area. Some, like the Caswells and

In this hand-copied page from the music book Thure Kumlien brought from Sweden, the music is written in two parts with a part for voice. Playing the flute and singing with his Swedish friends helped Kumlien get through some hard years in Wisconsin.
WHI IMAGE ID 147778

Ogdens, settled on the south shore of Lake Koshkonong in the 1830s hoping to cash in on the lake and river transportation boom that never happened due to the cancellation of the Milwaukee and Rock River Canal. Though there is no indication that Kumlien knew Caswell or Ogden, he had dealings with men he called "the English boys" who had a farm nearby. Taking a big step in July and probably going into debt, Kumlien bought a team of oxen from these English boys. The Devoes, who later played an important part in Mellberg's life, had arrived in the area from Michigan in the early 1840s, before the Swedes. They were neighbors Kumlien mentioned in his journal, as was the Samuel Kirby family from Connecticut.

When Gustaf Mellberg married Juliette Devoe in 1846, this cabin built by Juliette's parents in 1842 became a place the Kumliens often visited. WHI IMAGE ID 147668

More important than people such as Chippert, McAdam, and "Bickel's wife," whom Kumlien mentioned once or twice each in 1844, were Elias Downing, his wife, Polly, and their children. The Downings, who had come from Ohio, had quickly befriended Thure, Christine, and Sophia— whom Kumlien often called "Fia." The lives of the Downings were intertwined for years with the lives of the Kumliens and of the Reuterskiolds when one of the Downing sons, Samuel, married Anna, a daughter of Carl and Maria Reuterskiold. Kumlien worked at the Downings', helping with the threshing, and in his journal, he noted that he and Elias "talk[ed] about plowing" and went hunting together. The women of the two families worked together and visited together. Christine arranged with Downing for "the pig," whatever that meant. During one afternoon visit with the Downings, the Downings gave Wallberg a loom shuttle. Wallberg also stayed at the Downings', whether as a friend or as hired help it is not clear to us and, perhaps, was not clear to them. When Kumlien was sick one day and couldn't go to Fort Atkinson, Downing brought the family the sugar and tobacco they needed. On March 4, when Polly and one of her daughters visited the Kumliens' home, Kumlien wrote a bit of rare praise: "Fine neighbors."

Another fine neighbor was Henry Charrick, whose name does not appear in any Koshkonong area census, history, or cemetery but does appear often in Kumlien's journal. Kumlien helped Charrick with threshing in exchange for five bushels of wheat and helped with plowing and other unnamed work. He loaned Charrick oxen several times, and it was for Charrick that he built a cupboard. The Kumliens often visited with the Charricks. They had "Tea at Charrick's," and Charrick brought them some of his new ale. They spent Midsommar at Charrick's, and Charrick put Kumlien in touch with his brother-in-law, in Jefferson, from whom Kumlien bought a chair.

The Kumliens lived in what became Sumner Township on the border between Jefferson and Dane Counties. Their land was on the eastern edge of what was called Koshkonong Prairie—a large Norwegian settlement. Norwegian farms of several hundred acres each were spread over parts of nine townships. More than five hundred Norwegian families, nearly three thousand people, settled there in the 1840s. Immigrating to America for very different reasons than the gentlemen Swedes, these Norwegians were experienced farmers who came for better farmland and bigger farms. First settling in northern Illinois or southern Wisconsin, they later moved north away from marshy land at Rock Prairie and Jefferson Prairie for "wood, water and good farming land."[7] For a time, Koshkonong Prairie in Dane County and part of Jefferson County was the largest settlement of Norwegian immigrants in America.

In 1844, Kumlien borrowed a measure of wheat from "Ole on the prairie," hired out his team of oxen to "Ole Norman," and helped "Ole Norwegian" stack hay. We don't know if this was three Oles or the same Ole, but apparently there was just one "Norwegian smith" to whom Kumlien took his broken plow.

From February to June 1844, Kumlien mentioned "Janson" as frequently as he did Downing, Mellberg, and Charrick. But in July of that year, "Janson" dropped out of the journal and "Johnson" appeared as frequently as Janson had. Either Janson changed his name to Johnson or Kumlien changed the spelling of his name in his journal. It's clear that this is

In this 1887 outline map of Jefferson County, Thure's property (marked "T Kumlien") appears just west of the Busseyville post office and near land belonging to Thomas North and Carl Hammarquist. WISCONSIN HISTORICAL SOCIETY MAPS COLLECTION

the same man, a neighboring Norwegian farmer from whom Kumlien bought meal, flour, and seed and borrowed tools. Kumlien helped Johnson construct his bridge over the creek. He helped Johnson thresh his wheat and, on four July days, helped him rake his hay. While Kumlien was still living in a shanty and collecting logs for a cabin, Johnson was past those stages and was building a frame house. Kumlien helped split shakes for Johnson's roof. He and Christine visited at Johnson's, and the two men hunted and fished together. In August, Johnson took some of Kumlien's wheat to a mill, and Kumlien gave Johnson a crock of sausage.

Within walking distance of the Kumlien cabin was the cabin of well-read Norwegian immigrant Bjorn Anderson and his upper-class Swedish wife, whom the Kumliens visited now and then. A Quaker from a peasant family, Bjorn settled first in the Illinois Fox River settlement, then came north to Albion Township on Koshkonong Prairie in 1840, bringing his wife and children the following year. Their son, Rasmus Bjorn Anderson, became even more important in Kumlien's life in the 1860s when he was instrumental in getting Kumlien a faculty position at Albion Academy. Rasmus Anderson became a nationally known Norwegian scholar and diplomat and the founder of the Scandinavian studies program at the University of Wisconsin in Madison—the first in the nation. He wrote in his memoir, "It was the delight of my boyhood to walk the three miles from my home to visit the home of Kumlien, which was filled with stuffed birds of all kinds. Once I brought him an owl that I had shot and he gave me a stuffed bluejay for it."[8]

Rasmus Anderson also noted in his memoir that in the early 1840s, "the Indians had a camp on my father's land."[9] Native people had been living on the land surrounding Lake Koshkonong for thousands of years. Kumlien's first mention of them in his journal was on March 20, 1844, when he reported that "an Indian was down in the creek fishing."[10] Since the 1820s, the Ho-Chunk and Potawatomi people of Koshkonong had been suffering a devastating series of forced removals by the US government. In treaty after treaty, the tribal nations were coerced and pressured into ceding their homelands and moving west to live on reservations. However, in 1844, some determined Native people were still living on the land near where Kumlien settled and over which he passed every day. Stories of the 1832 Black Hawk War in northern Illinois and southern Wisconsin must have passed among the white settlers of Koshkonong country as the farmers met Ho-Chunk and Potawatomi people fishing in the lake and creek and plowed up Indian mounds. In 1837, settler George W. Ogden had recorded in his journal that "always friendly" Indians camped on the north side of the lake "scouting about the river and lake here, fishing and hunting every day." The Rock River, Ogden said, was lined with their canoes, and many nights the settlers heard the Indians' drums. Ogden also reported seeing Native people skillfully hunting: "We saw one shoot a duck with a bow and arrow. He did it handsomely."[11]

Kumlien was a hunter and apparently a good one. He spent more time than most settlers in the woods hunting for meat for the table, for furs to trade and sell, and for birds to preserve and collect. He was also a thoughtful and observant man who would have wondered about the people who had been hunting this land for thousands of years before him. As he hunted birds and squirrels, trapped muskrat, and fished in the creek and the lake, or as he followed the old path from the spring down to the lake shore, Kumlien was moving as generations of Native people had moved from one habitat to another—from mixed woodland down through marshy low-lands where the grass was five and six feet high along Koshkonong Creek to the oak savanna bluff and an ancient Oneota village site north of Kum-lien's property above the wild rice–covered lake teeming with fish and waterfowl. Carcajou Point, where Thure and Christine had spent their first winter with the Downings, was the site of White Crow's Ho-Chunk village—at one time the largest Native village on the lake.

A few years later, when early Wisconsin scientist Increase Lapham surveyed Indian mounds around Koshkonong, he wrote that because the area was so abundantly supplied with the rice and fish the Native people relied on, "we were not surprised to find numerous traces of Indians on the banks of the lake, which are known to have been occupied until a very recent period."[12] Kumlien, too, would have found many indicators that the area had quite recently been occupied by Native communities. Most of the paths and roads he walked upon had first been Indian trails. As he plowed, a projectile point or arrow head might have turned up in the black loam. Or at the Carcajou Point Ho-Chunk village site, he might have scuffed up shards of broken pots. After a heavy rain, a stone axe might have washed out of the creek bank. Or he could have spied the dull gleam of a copper spear point in the shallows of the lake. Some of these objects were ancient reminders that people had lived here for thousands of years, and some were reminders of the thriving community that the US government was brutally forcing from its land at the time Kumlien was settling there.[13] Kumlien made drawings and tracings of some of these items, and later in his life, he donated some of what he had collected to the Milwaukee Public Museum. Over the years, thousands of artifacts have been found in the area by farmers and collectors and archeologists.

On Bjorkander's forty acres, just north of Kumlien's forty, a group of twenty-seven Indian mounds lined the crest of the ridge—two very low conical mounds, four oval mounds, three linear mounds, an effigy mound, and seventeen conical mounds. Later, when Bjorkander died in 1851, Kum-lien bought those forty acres and preserved the mounds.

When Kumlien stood on his ridge and looked toward the lake and the low blue hills to the southeast, he was standing on a place important to the Native people who had built the mounds. The mounds must have brought to Kumlien's mind the great mounds at Gamla Uppsala—Old Uppsala.

While a student, Kumlien had been only a few miles from three famous burial mounds, called then the Royal Mounds, which had been protected by custom and law for perhaps a thousand years.[14] Kumlien, who had read the old Scandinavian stories, retold the sagas, recited the poems, and sung the old songs, could picture the Uppsala mound builders.[15] He must have wondered about the people who built these North American mounds.

Though thousands of Indian mounds were destroyed by plows at the time of white settlement, Wisconsin remains the epicenter of effigy mounds in North America. Yet, unlike many farmers who plowed up or otherwise destroyed mounds on their property—after they had plundered them for artifacts—Bjorkander and Kumlien left the mounds intact on those forty acres.[16] At a time when there was monetary value in planting all the acres one could, both Bjorkander and Kumlien must have understood the higher value of keeping these sacred sites intact. They still exist and are now called the Kumlien Mounds.

The American Indians' seasonal rounds among the wetlands—established thousands of years earlier, when glacial ice retreated—had been irreparably broken. In the 1840s, Kumlien didn't see the tribal nations in their strength.[17] He knew the Native people as refugees. As Ho-Chunk historian Amy Lonetree explains, throughout these decades of forced removals, "groups of our ancestors kept returning to Wisconsin, even though it meant living as 'fugitives' in our homelands."[18] Kumlien could not have seen even the little of the Ho-Chunk Nation that Juliette Kinzie had seen on the Rock River in 1831: "a collection of neat bark wigwams, with extensive fields on each side of corn, beans, and squashes, recently planted, but already giving promise of a fine crop."[19] Kumlien and the others who arrived in the 1840s did not see the Native people when they came from smaller winter camps to the great breadbasket of Koshkonong in the spring to plant corn, beans, and squash; to snare, net, and shoot with bow and arrow ducks and geese; and to gather mussels and duck eggs. In the fall, they gathered more water fowl and harvested the wild rice that covered the lake. Winnowed and cleaned and dried, stored rice would keep people alive all winter. By the time Kumlien came to Koshkonong, both the Ho-Chunk and the Potawatomi had lost their ancestral lands, and in many cases their lives, as a result of the federal government's policies of colonization. Though some determined Native people still remained in 1844, Kumlien saw mostly the ruins of bark-covered wigwams and abandoned canoes.

But when he walked down to the lake from his cabin, past the spring above the creek and the marsh, Kumlien was walking on a path used by many people before him, a path between the sacred ground at the mounds to the village on Carcajou Point.

7

THE STARVING TIME

1845–1849

By the beginning of 1845, the splendor of the Koshkonong landscape may not have been foremost in Thure Kumlien's mind. His muscles and callouses had hardened from wrestling a living from forty acres, and when Kumlien walked out on his land, where he had once seen only beauty and birds to collect, he must have now seen stumps to grub out, wood to cut and split, and land to plow.

When the harsh seasons and the struggle to feed their families became an inescapable reality to the Swedish settlers, most would put their heads down and work—some, like Kumlien, with growing reluctance. A few, like Gustaf Mellberg, flourished as farmers. Some tried moving to other parts of the country. Some gave up farming for other work. Some did not survive.

The years between 1845 and 1850 were the hard trial years for the Swedes in Wisconsin. In his work journal, Kumlien recorded the difficulties of this period at Koshkonong. Gustaf Unonius, whose life reconnected with Kumlien's in 1846, vividly recorded his starving time at Pine Lake.

During these challenging years, as Kumlien and Unonius realized they were ill suited for the manual labor of farming, they were sustained by the friendship of their fellow Swedish immigrants until eventually their intellectual interests and abilities offered them other possibilities.

One late afternoon in the summer of 1842, Unonius was fishing out on Pine Lake after a long day of cutting hay in the swamp meadow. From his

This early print shows the first log cabin and two outbuildings built by Gustaf Unonius on the shore of Pine Lake at "New Upsala." PUBLIC DOMAIN, GOOGLE DIGITIZED

little boat, he watched a man come riding up to his log cabin. He heard the man speak in Swedish to his wife, who turned and waved for Unonius to come in. When he stepped ashore and recognized the young man as a popular officer at Stockholm balls, his heart sank. He assumed, correctly, that the young military man had come to the Wisconsin Territory after reading Unonius's letters in *Aftonbladet*—a decision Unonius feared the man would soon come to regret. Unonius's own experience of less than a year in the Wisconsin Territory had already taught him that the settler's life might be suited to the laborer or artisan, but it had "nothing to offer young officials, military officers, and poor students."[1]

After Unonius's initial excitement at the beauties and opportunities of the land he had selected on the west side of Pine Lake, the hardships of the settler's life had darkened the bright promise of this new country. With kinsman Carl Groth, Unonius had cut the logs for a one-room cabin, which was quickly raised with the help of twenty-three neighbors and friends. Unonius and his wife, Margareta, along with Christine Södergren (Margareta's maid and friend) and Carl Groth, moved into the cabin. As pretty as the cabin was in their eyes, with only grass stuffed between the logs, it was so cold that a pan of milk left on the table would freeze overnight. Still, this was a welcoming home where new immigrants and settlers gathered, and Unonius and his wife gave shelter to as many as ten people at a time, with the women sleeping on the main floor and the men sleeping in the attic under a leaking roof. Before the birth of Gustaf and Margareta's first child in November 1842,[2] an extension was added to the cabin, giving the mother and baby boy some privacy.

During their second winter at Pine Lake, the severe winter of 1842–1843, some of their small herd of cattle died of hunger or froze to death. In the early spring, called the "starving time," when the winter's store of food had given out and there were no greens or grains for the animals, their oxen barely had the strength to work. Late that winter, the burden of hunger for Unonius and the members of his household was almost unbearable. In desperation, he and a neighbor plowed the previous year's potato field, gathering the few frozen potatoes that had been missed in the fall. In that same winter of 1842–1843, a Swedish neighbor's infant died and the first grave was dug at Pine Lake.

Between 1842 and 1845, a few more Swedes, Danes, and about fifty Norwegian families settled around Pine Lake. In their home countries, most of these Scandinavians had belonged to Lutheran churches. But, as there were no Lutheran congregations or ministers in the area, the families gathered in cabins for worship, often in Unonius's cabin. In January 1842, the Swedes were visited by a group of three young Protestant Episcopal missionaries who spoke of the Episcopalian Protestant Church as a sister church to the Lutheran Church of Sweden, founded on the same principles during the Reformation. One of these missionaries, Reverend James Lloyd Breck, who returned several times that year, impressed and befriended Unonius, giving him a Book of Common Prayer. Unonius and Breck talked often, and Breck conducted several Episcopal services that year for several dozen of the Lutheran settlers. On December 5, 1842, Reverend Breck baptized Gustaf and Margareta's first son. Not long after this, the three missionaries settled three miles southwest of Pine Lake, where they founded a school and a seminary at Nashotah Lake. Still without their own minister, the Scandinavian settlers continued to meet in various homes every other Sunday, with the sermon prepared by the host.

On May 7, 1843, Unonius hosted the service and delivered a sermon, after which Pastor Breck conducted holy communion for twelve Swedes and four Norwegians and offered them the services of his Episcopalian church. Some of the settlers joined the Episcopal Church, and Unonius continued to follow the liturgy of the Church of Sweden. In September 1843, a group of the Pine Lake Swedes came to Unonius and suggested that he be ordained in the Protestant Episcopal Church so that they would have a religious leader and a teacher for their children, since there seemed to be no prospects for a Lutheran congregation. Unonius, they said, understood their language and customs; he would maintain the spirit of Christianity among them. The first Episcopalian church in the area was completed at Nashotah in November 1843.

By that time, after struggling for nearly three years, Unonius knew that he had no future as a farmer. So, in January 1844, though he had reservations about joining the Episcopal Church, he began to study for the Episcopal ministry. He and his wife rented out the cabin they had built and the land they had cleared. Margareta and their child stayed in a borrowed cabin during the weeks Unonius stayed at nearby Nashotah studying for the ministry. Because of his university background, his study took less than a year. Unonius was ordained in the fall of 1845, quickly becoming pastor at Pine Lake. One of his earliest and saddest ministerial duties was the burial of their friend Christine Södergren, who had come with them from Sweden. Christine died in childbirth. The following May, Unonius also buried his and Margareta's first son.

Over the next four years, Unonius's ministerial work would take him to Swedish and Norwegian settlements in the east and southeast of the Wisconsin Territory, including Koshkonong. Unonius would struggle financially as a minister, as he had as a farmer, but he had rightly decided to make his way in the world using his mind, education, and social skills rather than relying on the sweat of his brow and the strength of his back.

On March 11, 1844, Kumlien wrote in his journal: "The girl was born 11:00 (about) by M. The girl weighed 8 #." This cryptic note records the birth of a daughter to Thure and Christine Kumlien. She weighed a healthy eight pounds. Twenty-three days later, on April 3, 1844, she was taken to Fort Atkinson to be baptized: "All went to Ft. Atkinson. Baby baptised as Christin Agusta Sophia (by Mellb.)" "All" must have been Thure and Christine, their baby, Sophia Wallberg, Mellberg, the Downings, and the Reuterskiolds. They probably rode in Downing's or

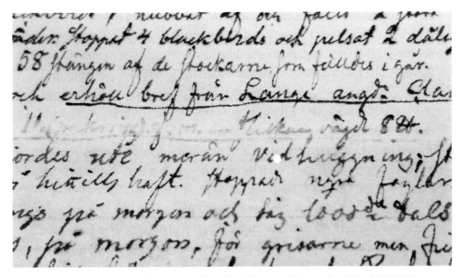

This faint, partially rubbed-out entry from Kumlien's work journal on March 11, 1844, includes information about his newborn daughter, Agusta. It reads: "11. The girl was born 11:00 (about) by M. The girl weighed 8 #." THURE L. KUMLIEN PAPERS, WIS MSS MQ, BOX 2

Reuterskiold's wagon pulled by oxen.[3] The Kumliens called their baby girl Agusta or Gusta.

By the fall of 1845 on Lake Koshkonong, Kumlien and Christine were sharing work and Sundays with friends and good neighbors. Their baby Agusta was learning to walk. Kumlien occasionally collected bird skins and sent boxes of specimens to museums.[4] In the beauty and abundance of this new land, the Kumliens experienced endless labor, unpredictable weather, and snakes in the house—a reality that bore little resemblance to the romantic dream of the cabin in the forest. Kumlien wrote: "The stable, the hog house, and the chicken house need repairs; there is no fence around the wheat bed. What is worse is the condition of the house. I have this year killed four small brown snakes that have crawled in between the walls."[5]

Despite the snakes, the early part of that year had been productive. The family had gone to the dedication of the new Norwegian church in Albion at the end of January. In January and February, Kumlien cut house logs and fence rails. In February, he saw a migrating flock of small gray birds he couldn't identify in the oaks next to his newly cleared field. He heard a blackbird, and he heard a crane. On a fine February day, Kumlien "fixed a prairie chicken and a wood cock,"[6] meaning he prepared the greater prairie-chicken and American woodcock as study skins or mounted them. In March, he labeled and packed more than thirty birds for two museums.[7] In April, he sold a muskrat skin, caught and sold several bushels of fish, and shot and sold more than fifty greater prairie-chickens, bringing in a little cash. Christine planted new peas and cucumbers. In May, Kumlien began spending time with Reverend Dietrichson of the new Norwegian church. A lawsuit resulting from Dietrichson's ouster of a member of his congregation was the talk of the community. In June, Kumlien planted and hoed, worked up new fields to plant, and helped his neighbors. In July, he and Christine and baby Agusta celebrated the Fourth of July "with Germans."[8] And the community celebrated the wedding of Anna Reuterskiold and Preston Downing. By the end of that month, Kumlien had three acres planted or ready to plant. He spent much of August cutting and stacking hay with Sven Bjorkander, who now lived on the adjoining farm to the north. But at the end of the month, Kumlien began to feel sick, and by the first of September, he and Christine, Sophia, Bjorkander, and Mellberg were all ill.[9] Because Kumlien was too sick to control his oxen, they wandered off. The garden froze on September 22. And after that, the year got steadily worse.

On October 1, 1845, Kumlien wrote, "My day of chills." He was sick with what was probably malaria. The fevers and chills, shaking, and weakness kept him in the house. On October 8, Kumlien wrote that Agusta was "sick from her teeth" and that five people in the neighborhood had chills.[10] An early hard freeze on October 15 damaged their cut hay, corn and pumpkins still in the field, and potatoes still in the ground. When she could leave her sick child and sick husband, Christine dug potatoes with Sophia, who came from where she was working nearby to help. The leaky house was cold. Kumlien wrote, "Shook all day was unable to work and went to bed."[11] A neighbor and friend whom Kumlien called "Farbro," or "Uncle," was taking his meals with the Kumliens. Apparently less sick than the Kumliens or their neighbors, Farbro was able to help with cutting wood, fetching water, bringing in potatoes, and finding the wandering oxen.

In the middle of October, it was clear to Thure and Christine that Agusta was seriously ill, though it's not clear to us what the baby's illness was. She may have had malaria, too. Dr. J. C. Dundass, thought to be the best doctor in the area, was called in, though Kumlien worried that they still owed him $5.25 for his help during an earlier illness of Christine's.

On October 23, Kumlien wrote in his journal that Agusta "had a heart attack which lasted until two o'clock in the night and which I fear will be the end of her."[12] Agusta Kumlien, a little over nineteen months old, died the next evening.[13]

The following morning, still sick with chills, Kumlien had to bring in the cut hay. Leaving Christine and probably Sophia to prepare Agusta's body for burial, he first had to go look for his infuriating oxen. That morning, perhaps as he was bringing the wandering animals home, he sold his oxen to a neighbor who offered him thirty-six dollars in gold and silver for them. That afternoon, Kumlien visited his friend Carl Reuterskiold, and the two men made a coffin for Agusta with two wide basswood boards Kumlien bought from Reuterskiold for twenty cents. A man named Staples helped them finish the coffin. Bjorkander worked all morning digging the grave in nearly frozen ground at what is now called Sweet Cemetery. Located about two miles from the Kumliens' home, just across the Dane County line, these two acres had been donated by farmer Freeborn Sweet in 1842 as a place for his neighbors to bury their dead. Agusta Kumlien's would be the third or fourth grave in the cemetery. Christine, Sophia, Mellberg, and perhaps other friends walked up to the nearby hill carrying the coffin in their arms. They buried Agusta under the bur oaks in the savanna on top of a round hill overlooking what is now called Sweet Lake.[14] Kumlien was too sick to go to the cemetery. He wrote that Mellberg "buried the child in true Swedish form."[15]

On the first day of 1846, Kumlien wrote, "Give O Jesus Peace and Good Fortune. . . . Severe weather. I accomplished nothing."[16] But that day, he had moved some of his mounted birds and bird skins and deer skins out of their cabin and into the cabin of Farbro, likely giving Christine room to prepare food for more boarders. Farbro was eating at the Kumliens' for a dollar a month. More boarders would mean more cash, and food to serve more boarders did not seem to be a problem. Kumlien shot ducks and geese. They raised pigs and kept one or two cows for milk and butter. Christine must have done more than her part in preserving and preparing food. But Kumlien continued to have chills. Reuterskiold was very ill, and Kumlien was a witness to Reuterskiold's signing of his will.[17]

On January 14, 1846, Unonius came to Koshkonong from Pine Lake to visit and stayed for three days. This was more than a social call for Unonius, who was now an Episcopalian minister. As far as we know, Kumlien and Unonius had not seen each other for five years, since Unonius had left Sweden in 1841. Kumlien said nothing in his journal about the visit except that Unonius was there, but Unonius reported it thus, years later, when he had returned to Sweden: "I met again an old acquaintance from Uppsala, a former student, Th. Kumlien, who, with his lovely young wife, had departed in search of his Arcadia in the American West. It was quite interesting to see how he divided his time between his farming and his scientific research. Necessity bound his hand to the plow and the hoe, while his natural inclination directed his thoughts to flowers, birds, and insects."[18] Unonius then provided a glimpse inside Kumlien's cramped cabin: "A fine herbarium and a well-filled ornithological cabinet, which was not very well arranged owing to lack of space, hinted that he was devoting himself more to scientific than to agricultural activities."[19]

Unonius noted, almost in passing, that the Swedes of Koshkonong seemed to be doing better than the Swedes at Pine Lake, many of whom were struggling even more than Unonius had. Most of the Pine Lake Swedes would eventually return to Sweden or move to Chicago. Unonius wrote that he remembered "with gratitude the happy hours" he spent with the Kumliens in Koshkonong. Though he couldn't visit the Kumliens as often as he would have liked because of the distance, he wrote that he "always enjoyed their hospitality in triple measure: as a minister, fellow countryman, and friend," and he experienced among them "a kindliness" that made his ministerial duties at Koshkonong seem a "refreshing rest."[20]

The other Swedes experienced the Kumlien home as warm as well. For a time, beginning in late January 1846, Carl Groth and Carl Hammarquist lived and took their meals with the Kumliens for a dollar per month each.[21] Bjorkander had moved into his new house nearby, but he was eating at the Kumliens. Though it would have been more work for Christine, the company of the Swedish men must have been welcome to the Kumliens on the long winter evenings.

Groth, who had come to Pine Lake with Unonius, had later left Pine Lake and traveled to New Orleans with Hammarquist. In New Orleans, they sold newspapers and tobacco and worked in a theater, both of them returning to Koshkonong with enough money to buy land. Hammarquist bought land near the Kumliens. Late in January 1846, they were all invited to a party at the Reuterskiolds', where they drank and danced and stayed through the night, though Reuterskiold was sick. In the early winter, Kumlien and the other Swedes cut timbers, hauled house logs and loads of firewood, and worked on the road near the sawmill dam. And on February 7, Kumlien wrote that "we Swedes" laid up a layer of logs for Mellberg's house. On February 11, "Uncle and others came and talked all day and played cards."[22]

Carl Hammarquist (who later changed his name to Charles) came to Pine Lake from Sweden in 1843, then moved to Sumner Township in 1846 where he farmed, married Josephine Reuterskiold, and fathered nine children. Hammarquist was the proprietor of a general store in Busseyville and the chairman of the Sumner town board, and in 1860 he was elected to the state legislature. WHI IMAGE ID 69870

In February, Kumlien took a rare trip away from Koshkonong to Pine Lake. He and Groth set off on foot, stopping for the night at Crowder's Inn at Rome. Kumlien spent a week with Unonius, and he also visited with other acquaintances from Sweden.

On February 28, Unonius again visited Koshkonong—not just as a friend but as an Episcopal minister. On March 1, he held a service with communion at the Reuterskiold home. After that service, Kumlien wrote that he "went in by writing for Unonius," meaning that he joined Unonius's Episcopal church.[23] Eight people joined Unonius's church that day: Carl and Maria Reuterskiold, their daughter Anna, Gustaf Mellberg, Carl Hammarquist, Sophia Wallberg, and Christine and Thure Kumlien. On March 2, Unonius left for the two-day walk back to Pine Lake.

Kumlien had temporarily escaped the grind of labor by visiting with old friends that winter, but he was behind in his work. He wrote, "Now is March gone and I have much undone—no new rails, no house logs except for long side of house that looks poor; seems too thin."[24] His more single-minded friend Mellberg had a house raising on April 2 and 3, though he and Kumlien were both sick.

The springtime caused Kumlien's mind and life to be even more divided. In April, he cut rails, laid up fences, and dug and planted for two days. Then he shot three greater prairie-chickens and "fixed" them, and he shot a sandhill crane and skinned it the next day. On April 30, in "lovely weather all day," he split fifty-two rails, but his mind was elsewhere: "O to now be in beautiful old Upsala!! Spring is come! Up and fly in the sweet air!"[25] Instead of flying in the sweet air, Kumlien plowed with Farbro's oxen, and he and Christine planted onions, spinach, tobacco, cucumbers, beets, melons, potatoes, and an acre of corn. On May 10, the couple visited "the Germans" and bought a dog from them for a dollar "to be paid for later in some way."[26]

There were celebrations in June 1846 when Unonius came to Koshkonong to perform the marriage ceremony of Gustaf Mellberg and his neighbor Juliette Devoe at the home of the Reuterskiolds. Then Kumlien celebrated the Swedish Midsommar Day by hunting in the morning and in the afternoon attending a gathering, again at the Reuterskiolds'.

While Kumlien was working on the farm and now and then collecting and preparing birds, Christine was spinning, milking, churning butter, looking after the pigs and the dairy cows, molding the candles, cleaning the shanty, and cooking for them and the two Swedes who were boarding there. Sometimes, Thure and Christine worked together in the fields hoeing corn, raking hay into cocks to dry, and harvesting bundles of wheat.

In August, Christine was sick for a week with what Kumlien called "bilious fever." Dr. Green, a neighbor who bought mounted birds from Kumlien now and then, was there to see her each of the seven days she was sick. Kumlien complained: "Because of Christine's illness I must look after meals and house so Uncle and Swen quit eating here for present." But his sister-in-law, Sophia, came one day and "fixed things up," then came back to help with the cooking and to look after Christine.[27]

The last of the 1846 celebrations took place in September in Pine Lake when Carl G. Hammarquist and Josephine Reuterskiold, who had met on the *Svea*, were married by Unonius at his home.[28]

In 1846, Kumlien's journal entries primarily consisted of notes about illnesses, farm work, the occasional bird shot, and celebrations. The entries for the following two years would look much the same, though with fewer celebrations.

In the cold dreary month of January 1847, six little pigs died of the cold, and wheat for flour was hard to come by. Kumlien fixed only two bird skins, but the Kumliens visited with the Hammarquists five times. In his journal entries, Kumlien lists "nothing in particular," "nothing out of the ordinary," "nothing today," and "nothing important, too cold."[29]

On February 10, 1847, Kumlien wrote in his journal that he visited Reuterskiold "who is again quite sick." Kumlien was not feeling well himself. On February 12, he "waked the night at Reuterskiold's," and on February 14, Thure and Christine were both at Reuterskiold's when Carl Reuterskiold died around midnight.[30]

The next day, Kumlien made a basswood coffin for his friend, and Sophia came home for the funeral, which was on February 17. Kumlien spent several days at probate court in Jefferson and assisted with the appraisal of Reuterskiold's property.

Of the Swedes who had traveled together on the *Svea* and settled at Koshkonong in 1843, Reuterskiold was the first adult to die, though the Reuterskiold's twins had died in 1843, and a child of Carl Reuterskiold's daughter Anna and Samuel Downing had recently died.

By the end of 1848, James Benneworth was farming near Koshkonong. The very well-educated Mellberg was now married to his neighbor Juliette Devoe, merging two farms. Sophia Wallberg was working in various people's houses. She could apparently do anything—care for children, cook, spin, care for livestock, work in the garden and the fields—so she always had work. And Kumlien was slowly accumulating logs with which to build the walls of a more substantial dwelling. When he had enough, he would call his neighbors together for a house raising, as they had called on him.

Kumlien's first years as a settler were years of hard physical labor with the occasional collecting and fixing of birds. Each of these areas of endeavor was diluted by the other, so that Kumlien's progress was slow in both farming and collecting. Yet, during those years he was successful at friendship.

In Kumlien's journal, Gustaf Mellberg is always referred to as "Mellberg," but Sven Bjorkander is "Sven." Though Bjorkander had come to North America two years earlier than the Swedes on the *Svea*, he became a close friend of the Swedes at Koshkonong. Born in Ekby, Västergötland, in 1818, Bjorkander attended school with Kumlien and Mellberg at Skara Gymnasium. He enrolled at Lund University in 1841 but left for America at the end of May 1842. By the middle of that September, Bjorkander was at Pine Lake with Unonius.

Later that fall, Bjorkander and several others were clearing land and building cabins while living in a brushwood hut when, as Unonius wrote, "Owing to carelessness and over-exertion, Bjorkander fell ill. His illness, serious in itself, was made still more so by the nature of the instruments used in a surgical operation performed on him."[31] Unonius thought that Bjorkander would surely die. For two and a half months, he was cared for in the attic of Unonius's cabin where skins and blankets hung around his bed to protect him from the wind and rain that came through the roof.

Once Bjorkander recovered, he was in debt to the doctor and to Unonius for food and care. And somehow he had almost no clothes. As soon as he was able, he began working. When his debts were finally paid in early October 1843, he bought new clothes, and he and Groth traveled to New Orleans.

While we don't know exactly what Bjorkander did in New Orleans, we do know that he worked at a theater, which paid well, and there he remained until July 1844. New Orleans in the 1840s was nothing like the Wisconsin frontier of the 1840s. It was a lively, well-established city with a thriving cultural life. Both black and white society attended Shakespeare productions, music and dance performances, the opera, and masked balls. The St. Charles Theatre had been established in 1835. Famous actors such as Junius Brutus Booth traveled by riverboat down the Ohio River and then the Mississippi River to perform in New Orleans.

Though New Orleans must have been attractive to Bjorkander, he came back to Koshkonong with Groth, arriving on August 12, 1844, with enough money to buy the forty acres just north of Kumlien.

For the next five years, Kumlien and Bjorkander spent full days and half days working together cutting Sven's house logs and putting on Bjorkander's roof. Bjorkander bought a team of red oxen, which Kumlien sometimes borrowed. They hauled timber. Bjorkander and his oxen helped Kumlien break with Bjorkander's breaking plow, and Kumlien returned the favor. They cut and stacked hay together. They hoed potatoes. Bjorkander and Kumlien also socialized together and occasionally hunted together. Bjorkander disappointed Kumlien one day by going to see his German friends instead of helping Kumlien.

On January 29, 1846, when Groth and Hammarquist arranged to eat and live at the Kumliens', Bjorkander, who had by then moved into his new house, arranged to take his meals at the Kumliens' with his Swedish friends. On Good Friday 1846, in what Kumlien describes as "the worst weather one could imagine" with "nothing doing outside," Bjorkander and Groth stayed at the Kumliens' all day sitting by the fire talking.[32]

Bjorkander apparently joined a nearby Mormon church on January 29, 1849. Kumlien's granddaughter Angie wrote that "it was Sven's joining up with the Mormons that finished his romance with Sophia Wallberg."[33] Unfortunately, we know nothing more of this relationship.

On April 2, 1849, Bjorkander wrote a letter home to Sweden telling about his past seven years. On April 19 of that same year, Kumlien wrote in his journal that "Bjorkander and Henry Charrick [bought] the old saw mill" down at Koshkonong Creek in what is now Busseyville. In November 1849, Bjorkander and Charrick celebrated the raising of a new building at the sawmill.[34] Sadly, on July 15, 1851, Sven Gabriel Bjorkander died, apparently after an accident at his sawmill.[35]

Kumlien was one of the bondsmen in the settling of Bjorkander's estate. Two years later, the Kumliens bought Bjorkander's forty acres, which joined their property on the north.[36] Though the following years were happier ones for the Kumliens, there must have been many times as they walked or worked Bjorkander's land when they remembered and mourned the death of their friend.

In these early and often dark years in Wisconsin, Kumlien must have been haunted by anguish, wondering if settlement in this new world was a wise choice. Yet, during those years, the labor shared with old and new friends, the illnesses and deaths not suffered alone, and the celebrations of Midsommar, births, and marriages—all of this tied Kumlien more closely to his friends and to Koshkonong. And in the midst of uncertainty and grief, more than most settlers, Thure Kumlien would have been heartened by the glories of his home in the Koshkonong landscape.

8

AMERICAN BIRDS AND BIRD BOOKS

1850–1851

One Sunday in May 1844, Thure Kumlien walked to Fort Atkinson to mail a letter to a Mr. Dole in Buffalo, New York, a man he had met on the journey to Milwaukee. Later the same week, on a rainy Saturday, Kumlien wrote in his journal that he "shipped birds."[1] The birds likely went to Benjamin Dole, who owned a dry goods store on Long Wharf in Buffalo.[2] We can imagine that Kumlien had wandered into Dole's store on the wharf while waiting for the boat around the lakes and, seeing there a few mounted birds, began a conversation with Dole, offering to send him mounted birds when he arrived in Wisconsin. This was apparently Kumlien's first shipment of Wisconsin birds to an American buyer. We don't have a list of the birds he sent, but perhaps they were red-winged blackbirds or the crane and tern he had shot and prepared at the end of March. It would be interesting to know if and how Kumlien identified them.

In Sweden, if Kumlien saw a bird he couldn't identify, he could consult any number of friends who studied and collected birds. He and his friends could also consult books on the birds of Sweden and, very likely, museum collections in Uppsala and Stockholm. At Koshkonong, however, Kumlien was the only naturalist. His self-described "greatest handicap" when he first arrived, according to Angie Kumlien Main, was that he had no American books of ornithology or botany. Using his European books, he "was always comparing the birds of this country with those of Europe."[3] In the expanding United States of that time, ornithologists such as Kumlien who worked in isolation relied on books to inform them of the species of birds they observed and collected. Kumlien wanted John James Audubon's *Ornithological Biography* and *Wilson's American Ornithology* by Alexander Wilson. Though Audubon's works would remain out of Kumlien's financial reach for all of his life, Thomas Mayo Brewer's revised edition of *Wilson's American Ornithology* and then Kumlien's correspondence with Brewer himself would become Kumlien's mainstays for identifying American birds. He apparently asked his friend Gustaf Unonius, who regularly traveled around southeast Wisconsin as a minister, to find him those two books.

On March 21, 1848, Kumlien noted in his journal that he went to Pastor Dietrichson's house to pick up the copy of *Wilson's Ornithology* that Unonius had sent to him. Unonius wrote to Kumlien, "You have received, I believe, the book from Pastor Dietrichson. It cost $3.00, which little sum I have paid out. . . . The other book [Audubon's *Ornithological Biography*] which you asked me to buy costs $30. So I thought it best not to buy it."[4] Kumlien said he raised the three dollars for *Wilson's Ornithology* by selling six bushels of potatoes and collecting $1.50 in fees from appraising his late friend Carl Reuterskiold's estate.[5]

Kumlien's edition of *Wilson's Ornithology* had been revised in 1840–1841 by Brewer, who would become Kumlien's connection to the North American world of natural history—particularly ornithology and oology, the study of eggs. The story of this 1840 volume is in part the story of early American ornithology.

Some American birds had been painted by British visitor Mark Catesby and described by Swedish visitor Pehr Kalm before 1800,[6] but the Scottish immigrant, artist, and poet Alexander Wilson (1766–1813) was the first to set out to describe and paint all of the birds of North America. After arriving in America in 1794, Wilson took a job at a school near Philadelphia and at the botanical garden of William Bartram, whose late father, John Bartram, was well known to European naturalists as the premier collector of American plants. William Bartram, who soon became a close friend of Alexander Wilson, had three years earlier, in 1791, written and published the widely acclaimed *Travels through North and South Carolina, Georgia, East and West Florida*. Important to naturalists and Romantic writers in America and Europe, Bartram's *Travels* is still in print. A Quaker, Bartram

Kumlien longed for an illustrated book on American birds like Alexander Wilson's *American Ornithology*. This Plate 70 from Volume VIII would have helped him identify male and female long-tailed ducks, wood ducks, green-winged teals, canvasbacks, redheads, and mallards. UNIVERSITY OF WISCONSIN DIGITAL COLLECTIONS

was, in the late eighteenth and early nineteenth centuries, recognized in Europe and the United States as America's foremost naturalist.[7] Wilson became an informal student of Bartram's, learning from him the names of birds and plants and how to be a better observer of the world they both loved. Friendship with Bartram gave Wilson access to Bartram's library and the libraries of the American Philosophical Society in Philadelphia. In 1806, Wilson took a position at an important Philadelphia publishing house and almost immediately approached the publisher with his idea for writing and illustrating a work on the birds of North America. The publisher agreed to this proposal—if Wilson could collect two hundred subscriptions. By 1808, the first volume of this project was complete.

Wilson would spend the rest of his life at this demanding and arduous work. He traveled the country to observe and collect birds. He drew and painted the birds, arranged for engravings, sold subscriptions, and hired and supervised colorists. One year when he did not have the money to hire colorists, Wilson himself colored more than thirty-two thousand bird illustrations. Also a poet, Wilson wrote the detailed, lucid, accurate descriptions of the birds in his monumental work. Though he was not trained as a scientist, he was the first American to consistently use Carl Linnaeus's system of naming. His attention to the habits of live birds and their nests, his first breeding census of birds, his ability to generate hypotheses based on his observations and then test those hypotheses—these and many other of Wilson's qualities and abilities made his book invaluable to those collecting and observing birds. This magnificent work was brought to completion by his friend George Ord after Wilson's death in 1813 at age forty-seven.

Wilson's American Ornithology, after being updated and expanded by Charles Lucien Bonaparte and then Thomas Nuttall, was again added to and republished by Brewer in a one-volume edition. This affordable 1840 edition included ornithological observations by Bonaparte, Nuttall, and John James Audubon, along with small black-and-white cuts of Wilson's original bird paintings. Brewer's sixty-four-page, closely printed synopsis at the end of the book added up-to-date names and identifications of birds. If a person wanted to study North American birds at this time, he would absolutely want Brewer's edition of Wilson's American Ornithology.

Settlers such as Kumlien were living in what was primarily a barter economy. Kumlien could go to the store and pay for his goods with eggs and butter[8]—but increasingly, he needed cash. For example, in January 1847, Kumlien walked to Fort Atkinson, where he bought twenty-two bushels of corn for forty dollars, two bushels of barley for forty cents, an apron for thirty-two cents, cheese for six cents, opium and alum for three cents each, and licorice and camphor for a shilling each.[9] Cash had to be used for shot, matches, plug tobacco, and smoking tobacco.[10] Many of the settlers, including Thure and Christine Kumlien, raised pigs and sold pork for cash, getting five cents a pound for pork in April 1847.[11] Occasionally, the Kumliens sold butter, but in small quantities, for ten cents a pound.

The sale of mounted birds and bird skins would bring Kumlien cash as well as the pleasures of the knowledge and craft involved in processing the birds. In 1847, along with his farm work, Kumlien was selling a few birds to neighbors—two sandhill cranes to Dr. Green for a dollar, four birds to an Englishman, and several to a Dr. Head. Kumlien sold a glass box of birds for $1.75. The total amount he received from selling birds in 1847 was $7.75—and that was a good year. In November of that year, he noted that he wrote to the New York taxidermist John G. Bell, who had sold mounted birds to Audubon. Kumlien likely offered to sell bird skins or mounted birds to Bell, but nothing came of this.[12] In May 1848, a month when he'd seen American white pelicans among nearby tamaracks, Kumlien sold in Janesville a glass box with several mounted birds for $1.50.[13] However, he needed to sell many more bird skins and mounted birds to make a difference in his family's income.

In spite of being cash poor, on March 20, 1849, Thure and Christine finally "called for a house raising."[14] For almost six years, they had been living in a rough log building that was little more than a shanty. For that time, Kumlien had been slowly collecting logs for walls and beams and other materials for a real cabin. He had dug and hauled stone for a foundation. Most of his friends had built their cabins several years earlier. But that March, Sven Bjorkander, Ole Lind, and Gustaf Mellberg helped Kumlien get the last of the logs and the foundation ready. Then, on March 22, friends and neighbors came together and raised the walls and roof of the Kumliens' cabin. In his journal, Kumlien wrote, "The house raising went well; heard nothing different."[15]

The walls and roof were up, but there was still much to do on the cabin before the couple could move in. Over the next several months, besides the usual work of planting, weeding, and harvesting; trading work with neighbors; and occasionally fixing birds, Kumlien had to shingle the roof of the new house and chink the walls with mud. The siding for the house arrived on September 28 and 29—"16 foot oak boards from 'the old saw mill.'"[16] And the house had to be furnished. On October 16, Kumlien borrowed a wagon and, taking along some "birds in glass, some butter and some cash," he bought a stove for $13 and stove pipes for $2.50.

On October 24—the day after Kumlien went to the county seat and, with no comment or fanfare, became a citizen of the United States—Mellberg helped him nail up lathe in the new house. After three days of help from a neighbor named Christianson, though the walls were not plastered nor the floor boards laid, Thure and Christine moved into their

On October 23, 1849, on this naturalization form from Wisconsin's days as a territory, Thure Kumlien swore to renounce forever any allegiance to any foreign prince or potentate, particularly to Carl Johan, King of Sweden. He was recommended by the "competent testimony" of his good friends Elias and Preston Downing. WHI IMAGE ID 147669

new house on November 1, 1849.[17] On Christmas day, Christianson made them a table, and the day after that, Kumlien made cupboards. On December 28, Kumlien put panes of glass in a window of his new house.

On January 4, 1850, Kumlien wrote that he helped Hammarquist all day with *his* house.

On January 5, he "worked on a swan." This is the last entry in the journal.

Kumlien had begun the journal in 1844 as a "Work Journal," using it to keep track of the work he did for others and that others did for him. Apparently, the journal stopped because that kind of tracking was no longer necessary. In early January 1850, Kumlien for a time gave up the idea of making his living as a farmer. He rented out his cleared land to a neighboring farmer. He and Christine stayed in their new house and kept their gardens and orchard, but they left the farming to someone better suited to it. "Farming has of late become a poor business here," he wrote to Brewer, "and [especially] with me anyhow not being much used to hard labor."[18]

Kumlien took a job early in 1850 at a sawmill in the village of Jefferson, where several sawmills on the Rock and Crawfish Rivers were doing a booming business as more land was cleared and more frame buildings were built. Kumlien most likely boarded in Jefferson for most of each week, walking the fourteen miles home at the end of the week to be with Christine. Kumlien's life was changing.

Wisconsin was changing, too. In 1850, Wisconsin had been a state for almost two years. The population of white settlers in Wisconsin was three times what it had been in the early 1840s. Jefferson County in 1840 had a population of 1,594, but by 1846, the population had tripled to 4,758.[19] Though some settlers left Wisconsin to search for gold in California in 1849 and 1850, increasing numbers of immigrants were arriving from the eastern states and Europe, especially from Germany. Many communities in the state now had enough young people and money to hire a teacher and start a school. Some small towns, including Jefferson, Janesville, and Fort Atkinson, had weekly newspapers by this time. The first train in Wisconsin ran from Milwaukee to Waukesha on February 25, 1851.

With statehood came a university—the University of Wisconsin in Madison, about forty miles northwest of where Kumlien lived. The first university class met on February 5, 1849. Two years later, the first university building, North Hall, opened, serving as a dorm and classroom building. That same year, the board of regents ordered the collection of Wisconsin's plants, animals, and minerals to be compiled in a "cabinet of natural history." Also in 1851, Peter Engelmann founded a German-English

Academy in Milwaukee, which began to gather and display natural history specimens.

Kumlien lived in Sumner Township in Jefferson County about a mile east of the Dane County line. This country north and west of Lake Koshkonong was informally called Koshkonong or Koshkonong Prairie until towns began to be incorporated. Less than a mile south of Kumlien's property, where the meandering Koshkonong Creek crossed the Military Road that is today's Highway 106, a dam and a sawmill had been built in the 1840s. A store was built, then a church. A school held classes in 1847. Still, it would be more than thirty years before this crossroads would be called Busseyville, as it is now, after English settler Thomas Bussey, who bought the mill.

Kumlien's sawmill job in Jefferson, perhaps at Darling & Kendall's mill, apparently lasted only for the winter of 1850. He was home in the spring to continue to collect birds, to put in his garden, and to tramp the woods and the shores of the big and little lakes.

In June 1850, a man named Charles Holt, an editor who owned half interest in the *Janesville (Wisconsin) Gazette*, visited the Kumliens at their new cabin. Holt had worked on the *New-York Tribune* for Horace Greeley and had reported on Wisconsin's first constitutional convention in 1846.[20] Holt must not have been a man easily impressed, and he must have expected this rustic cabin to hold what most settlers' cabins held. So he was surprised when he opened the Kumliens' door to find beautifully mounted birds and insects, pressed plants, books on natural history, and poetry in several languages in this out-of-the-way place. He was surprised by the very well-educated Kumlien, "an individual so well informed and conversant with every department of Natural History."[21] Even more surprising to Holt was that Kumlien lived only sixteen miles from Janesville and Holt had never heard of him.

Holt went back to Janesville and wrote about Kumlien in the June 13, 1850, issue of the *Janesville Gazette*: "If any of our citizens desire the services or acquaintance of an intelligent naturalist, they will find such an one in Mr. Thure Kumlien, living near the foot of Lake Koshkonong. . . . The localities about his residence give him rare opportunities for procuring specimens, and they are prepared with a great deal of care and good taste. He has now ready for sale several cases of beautiful birds and we recommend those who may wish them to make an application to him."[22]

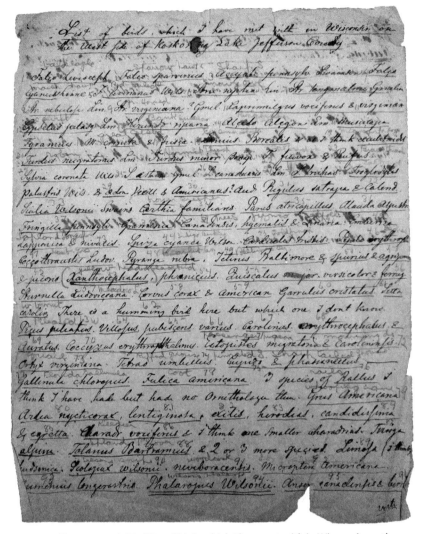

Thure Kumlien prepared this "List of birds which I have met with in Wisconsin on the West side of Koshkonong Lake, Jefferson County" to send to Thomas Mayo Brewer at the beginning of their correspondence. THURE L. KUMLIEN PAPERS, WIS MSS MQ, BOX 1

Holt had noticed not only Kumlien's work and abilities but also how well situated he was to study birds and plants.

Holt and Kumlien must have spoken about Kumlien selling bird specimens, because a few days after Holt's visit, Kumlien sat down with his *Wilson's Ornithology* and his notes on birds and prepared a list of 116 birds he had "met with in Wisconsin on the west side of Koshkonong Lake in

Jefferson county."[23] At the end, Kumlien wrote, "This list, of course, is very imperfect, but having not until late been able to get a book on the subject and little time to spend on hunting, it is very likely that I have not mentioned half of the birds we have here."[24]

The list began with the bald eagle and ended with the black tern and the common loon. In between are common birds such as the eastern phoebe, American robin, gray catbird, house wren, seven types of woodpecker, and birds that are now either uncommon or extinct, such as the snowy owl, passenger pigeon, greater prairie-chicken, and whooping crane.

A few weeks later, Wisconsin's first prominent scientist, Increase Allen Lapham, passed within a few miles of Kumlien's home on his survey of the Indian mounds of Wisconsin.[25] Traveling down Koshkonong Creek, Lapham wrote, "Koshkonong Creek appears not to have been a favorite place for the ancient mound builders at least we could not find any traces of their works."[26] If someone had told him of the twenty-seven Indian mounds on Kumlien's ridge overlooking Koshkonong Creek, he most certainly would have made his way to Kumlien's cabin. Leaving Jefferson at dinner time on July 4, Lapham and his nephew John spent the night at an inn in Cambridge about two miles from Kumlien's house; they were served roast pig and watched "a very pretty exhibition of fire works considering the newness of the country and the small-ness of the place."[27] The next day, they went to nearby Clinton, where they saw no "ancient works" along the banks of Koshkonong Creek, and since there was no shoe on the hind foot of Billy, the horse who pulled their wagon, they left Billy at a blacksmith's while they examined an outcrop of sandstone and limestone.

If Lapham and Kumlien had met at this point, what a happy meeting that would have been. Lapham would have been the first naturalist Kumlien had met in America. And Lapham would have been even more intrigued than Holt had been with Kumlien's collection of birds and bugs and plants.

However, Kumlien was about to make an important connection with a naturalist from Boston that would continue for thirty years.

9

THOMAS BREWER AND THE YELLOW-HEADED BLACKBIRD

1851–1853

In July of 1850, enclosed in a letter from Janesville newspaper man Charles Holt, Thure Kumlien received a letter from Boston naturalist Thomas Mayo Brewer asking if Kumlien would be willing to aid him in his endeavors to illustrate the eggs of North American birds.[1] Kumlien had bird skins and mounted birds ready to sell, but probably no eggs. Yet, he was so hesitant to write in English that he took almost a year to respond to Brewer's letter, finally writing him at the beginning of the following April.[2] Brewer wrote back in November of 1851 saying that a well-informed and isolated naturalist like Kumlien would be "a rare and invaluable aid to a kindred spirit."[3] Brewer was right. They were kindred spirits in their backgrounds and in their love of birds. And though they would never meet, the two men would be invaluable to each other for almost three decades.

Brewer was born in Boston in 1814 to a wealthy family. Educated at Harvard as an undergraduate and then at Harvard Medical School, he practiced medicine for a short time, then became a contributor to and publisher of the *Boston Atlas* newspaper and a partner in the printing firm of Brewer & Tileston. Brewer was a friend of John James Audubon late in the great painter's life. Audubon named a duck, a blackbird, and a mole after Brewer.[4] Kumlien knew of Brewer as the compiler of his edition of

Wilson's American Ornithology, the one Unonius obtained for him in 1848, the book that made it possible for Kumlien to identify American birds.

It was Kumlien's reluctance to write in English that delayed his response to Brewer. He wrote:

> Dear Sir,
>
> Not being enough acquainted with the English language to write the same any way proper, it is with some hesitation I undertake to write to you, but if I can make my writing intelligible then my object is gained and I thrust [trust] to your kindness to have forbearance with my stilistical [*sic*] faults. I am a native of Sweden and emigrated to America in 1843, but have ever since been shut up in the woods and my time most wholly devoted to hard work. I have been absent from home this last 6 months, but have now returned and as far as my time will permit I will collect eggs for you this season.[5]

Though Kumlien had been in this country for about eight years, his use of English would have been limited to conversations with Yankee friends such as the Benneworths, Downings, and Norths, who had settled on neighboring land in 1847; interactions at work in the furniture factory in Jefferson; and dealings at stores in Fort Atkinson and Janesville. He read his *Wilson's Ornithology* and very likely English newspapers such as the *Janesville Gazette*. But he would have spoken Swedish at home and with his friends Mellberg, the Reuterskiolds, Unonius, Hammarquist, and Bjorkander. He likely spoke Norwegian with his Norwegian neighbors to the west on the Koshkonong Prairie. And he may have used his German with the increasing numbers of German immigrants to the north. It's not surprising, then, that he was unsure of his written English. However, in a year or so, the English in his letter drafts would become perfectly acceptable.

While Kumlien was concerned about his pay for collecting for Brewer, neither of them was very direct about money. Brewer wrote that he knew Kumlien would "require some compensation" for his trouble and "for enlisting the boys in your neighborhood . . . to point out to you the nests they might discover." He said he would pay Kumlien "trusting that you will not place your valuable aide beyond my reach."[6] Brewer mentioned no figures in this letter or the next, and though he asked Kumlien several times what he charged, Kumlien finally just suggested that Brewer pay him what he thought right.

Both Brewer and Kumlien were clearer and more explicit about the birds and bird eggs. Brewer was even explicit about the bird eggs he did *not* need, sending a list of these, as well as another list of eggs he didn't particularly want but which would be good to have for trading with foreign collectors. And because he was not "conversant with the species which breed in your neighborhood," Brewer told Kumlien, "I would be very glad if you would at your leisure send me a list of them."[7]

Kumlien had prepared a list in the spring before Brewer even asked: "Perhaps I can get some of the following," then listed the sparrow hawk, whippoorwill, chimney swallow, kingbird, white-rumped shrike or great grey shrike, some warblers, indigo bunting, summer tanager, oriole, yellow-headed blackbird, white-breasted nuthatch, hummingbird, hairy woodpecker, downy woodpecker, black-billed cuckoo, passenger pigeon, pinneated grouse, sharp-tailed grouse, American coot, rails, whooping crane, bittern, great blue heron, American egret, killdeer, upland plover, Wilson's snipe, long-billed curlew, Canada goose, mallard, wood duck, green-winged teal, blue-winged teal, common tern, gull, and pied-billed grebe.[8] And this, Kumlien noted, wasn't even a complete list of what he had seen since 1843.[9]

He continued: "I have some of these and I know where most likely I can find some of the rarest and though it certainly will be a tiresome and rather uncertain hunting, especially among the grass [wild rice] in Lake Koshkonong, I will endeavor to get as many kinds as I possibly can."[10] Collecting the eggs of birds like the yellow-headed blackbird, American coot, American bittern, Canada goose, and some of the ducks that nested in the marsh would mean borrowing a boat or wading in the mud, fighting mosquitos, and keeping an eye out for snakes. He asked if Brewer wanted the nests of the smaller birds or only a description. For hawks' eggs, Kumlien would have to climb trees or hire boys to do it for him.

Kumlien ended this first letter to Brewer by promising to send a complete catalog of the birds he had seen at Koshkonong. And he explained that he had only had an American "ornithologie (your edition of Wilson)" for two years. He mentioned that he did have European works on European birds, among them Sven Nilsson's *Scandinavian Fauna*, either the 1817 or

1821 edition, written in Latin. He then asked Brewer what he had asked Unonius several years earlier: "Is there no small edition of Audubons work? Please to inform me about the price thereof."[11]

Kumlien wrote again in November 1851 and sent Brewer a box of eggs, which were received in late November. "I am at present working in the village of Jefferson," Kumlien wrote, "but being in hopes that my services will prove more useful to you, I think I will move down to Koshkonong in the Spring."[12] Kumlien was again working at the turning lathe in the furniture factory, receiving "good wages turning out table legs, bedposts, rollers, neck yokes, kegs, sets of tubs, handles, desk legs, and the like."[13] The work required manual dexterity, a good eye, and an aesthetic sense. Kumlien was apparently good at the work and enjoyed it—so much so that "later he bought a used turning lathe and made maple and black walnut furniture for Christine and their cabin."[14]

Brewer was pleased with the contents of this first box from Kumlien: "They were all acceptable and most of all the shore-lark's [horned lark's] eggs—I had received one before obtained on the coast of Labrador and the four you sent me were very acceptable. So too were the eggs of the whooping crane."[15] Brewer wanted more of all of the eggs Kumlien sent: "the quail's [northern bobwhite's] eggs and those of the red-winged blackbird, though, common here also, are always good in making exchanges with European collectors."[16]

He asked about other species. "Do not the dusky duck, or black duck as it is also called, and obscura and the green winged teal breed with you?"[17] He was sorry Kumlien was not able to find a nest of the black tern. Brewer thought of more and more birds, their eggs and nests and then skins as well, which Kumlien might send him, including the very rare short-billed marsh wren [sedge wren].[18] "Our society would like the skin of one [sedge wren], if you will procure us one," Brewer wrote, "and perhaps when I get to Washington I may find that some others of my friends would like specimens also."[19]

"Our society" was the Boston Society of Natural History, of which Brewer had been a contributing member since 1835. His "friends in Washington" were Spencer Fullerton Baird and Robert Ridgway, ornithologists at the Smithsonian Institution. Brewer, Baird, and Ridgway were soon to begin work on their three-volume *A History of North American Birds.*

While Kumlien and Thomas Mayo Brewer, pictured here in a portrait painted by James Woodhouse Audubon in 1838, kept up a friendly scientific correspondence for more than thirty years, they never met in person. *THE BOSTON SOCIETY OF NATURAL HISTORY, 1830–1930* VIA BIODIVERSITY HERITAGE LIBRARY

And the unidentified bird skin Kumlien sent along—a shrike, which Brewer thought might be the loggerhead—Brewer would take to his "friend Mr. Cassin, who will be able to determine more certainly."[20]

Brewer's friend Mr. Cassin was John Cassin (1813–1869), an ornithologist at the Academy of Natural Sciences in Philadelphia. Cassin was the most respected American bird taxonomist of his day. Five birds have been named after Cassin, though he mostly sat at his desk in Philadelphia studying trays of bird skins and only discovered one new bird species in his lifetime (the Philadelphia vireo, which Kumlien saw at Koshkonong). Cassin was at that time the ultimate bird identification authority, the man who was asked the thorniest bird questions. At the well-funded Philadelphia Academy, Cassin worked closely with Spencer Baird at the Smithsonian.

Kumlien, in his close reading of Brewer's edition of *Wilson's Ornithology*, would have realized that this new connection to Thomas Mayo Brewer had connected him also to the most important ornithologists of the Boston, Philadelphia, and Washington circles of naturalists. Kumlien was one of the many correspondents around the country who supplied Brewer, Baird, and other ornithologists in the east with bird skins, eggs, and nests. Kumlien would never meet any of these men in the east, but they respected and accepted and used his work for the rest of his life.

Brewer wrote that the whooping crane eggs had arrived safely and he would be glad to have more. And he wanted to know right away about the cranes of Wisconsin. Was the sandhill crane different from the white species? Did it have a white phase in Wisconsin? Brewer was trying to work out the "life histories" of the whooping crane, the white heron, and the sandhill crane. Brewer said he "should like very well to know more about the sandhill crane and if it is different from the white species."[21]

Kumlien had seen sandhill cranes and what he recorded as whooping cranes at Koshkonong since he had arrived. Early in the journal, he didn't distinguish between them, calling them simply *trana*, Swedish for crane. The cranes were shy birds and would fly off before Kumlien could come within shooting distance. In his everyday life, Kumlien used the cranes' common names when talking about them with farmers, who would bring him cranes to stuff, but in his scientific work with Brewer, Kumlien was careful to use the Latin binomials in *Wilson's Ornithology*.[22]

At this early stage of their communication, Kumlien was a conservative scientific observer, not often venturing to make definite identifications, but rather letting Brewer come to his own conclusions. In one letter, Kumlien wrote,

If, what here is commonly called "Sandhill Crane" should become white, it perhaps would be most likely to be the case with old birds or adult ones, birds who have for at least 6 years had nest and raised

young ones, or there would likely be at least one white one among
all the Sandhill Cranes I have occasion to see every spring summer
and autumn, but yet I have not seen or heard of any white Sandhill
Cranes. What they here call White Crane is the *Ardea egretta* [great
egret] which sometimes in August and September fishes along the
lake shore though all of them I ever have seen had nothing of the crest
or tuft of feathers hanging down behind.[23]

In the increasingly frank and congenial letters between Brewer and
Kumlien, we get a sense of what a pleasure it was for Kumlien to share what
he found in the woods and the lakes and the swamps and the marshes with
Thomas Brewer. When Brewer wrote that "our society" would like the skin
of a crane,[24] Kumlien wrote back:

> I have lived on the same place for nearly eight years and every spring
> there has been a pair of cranes on the marsh below my house; they
> have had their nest there and one spring they had it placed so I could
> see her sitting on her eggs from my window. I did not disturb her, as I
> loved to have the stately bird sitting on the marsh unmolested, but
> one of my neighbors had a different taste—set his dogs on them and
> fired at them without any other effect than that they have been rather
> shy since and keep on another side of the marsh.[25]

Though Kumlien "loved to have the stately bird sitting on the marsh
unmolested," he wrote to Brewer, "I will endeavor to get you a good skin
of one next spring. If I cannot shoot any myself, having no rifle, I will hire
one shot"[26]—this he was willing to do for the sake of the science and for
his friend.

At that time, the primary way to further the scientific knowledge of
birds was to collect their eggs and skins. Brewer suggested that if Kumlien
could shoot a bird on the nest with its eggs, they could be very sure of its
identification. The men in the east who depended on men in the field for
their specimens had to have a good deal of trust in them.

For another season, until the late spring of 1852, Kumlien continued
to work at the turning mill in Jefferson, earning what he called good
wages. Though the factory work would curtail his collecting of eggs during

This watercolor from John James Audubon's *Birds of America* features (from top to bottom) a tricolored blackbird, three yellow-headed blackbirds, and a Bullock's oriole.
NATIONAL AUDUBON SOCIETY

the nesting season, he had written Brewer in January that he could probably get the skins of some rare (or, at least, rare to Brewer) birds that year—Wilson's phalarope, the American white pelican, the black tern, and the yellow-headed blackbird. How much was the Boston Natural History Society willing to pay for the skins, he asked Brewer. Kumlien said he had to get enough money to cover his expenses since his "circumstances" were "rather narrow."[27]

Of particular interest to both Kumlien and Brewer was the yellow-headed blackbird—a black bird with a yellow head that arrived in the marshes around Lake Koshkonong in flocks in April. This bird reminded Kumlien of red-winged blackbirds—the birds divided into separate flocks of males and females, as red-winged blackbirds did, and they exhibited some of the same racket and flashes of color. But these birds were bigger, flashier, louder, yellower. Like the red-winged birds, flocks of yellow-headed males noisily staked out territory in cattail marshes, and the drabber females would arrive a bit later.

Though there is no record of Kumlien seeing the yellow-headed blackbird before 1850, he most certainly noticed this aggressive, raucous, beautiful bird. Kumlien, with his musical ear, would have noticed early on the harsh, almost comical, buzzing calls of the male. But as he had seen nothing like it in Sweden and this North American bird, of course, wasn't described in his Nilsson's *Scandinavian Fauna*, if he did see it in 1844, he wouldn't have known what to call it other than the common name the earlier English-speaking settlers used—yellow-headed blackbird. When in 1848 he got *Wilson's American Ornithology* and pored over the descriptions, Kumlien would have found the yellow-headed blackbird in Brewer's synopsis under "Subgenus Agelaius," blackbird—*Icterus xanthocephalus* [*Xanthocephalus xanthocephalus* in modern usage], or saffron-headed blackbird. He would have read: "Head, neck, and breast, orange yellow; two black bands over eye; rest of plumage, glossy black, except two white bands on wings. Female, principally of a chocolate brown color. *Male*, 9. Habitat, California and Fur Countries."[28]

Kumlien might have hesitated when he read that the bird was found primarily in "California and Fur Countries." Still, he knew what he had seen. This bird was the right size with the right white patches on the wings. After referring to *Wilson's Ornithology*, he would have realized there was nothing else it could be. *Wilson's* would have given him a name for the nameless bird. Kumlien had seen it in the marsh during its breeding season, so in 1850, he put it on his first list of the birds and bird eggs that he might be able to get for Brewer. And he wrote Brewer on March 30, 1851, "Perhaps I can get some of the following . . . Xanthocephalus."[29]

But by the end of 1851, Kumlien had not yet found nesting yellow-headed blackbirds, so he could not send Brewer nests or eggs that year.[30]

He wrote again in 1852 that "icterus xanth." was one of the rare birds he expected to find. But that year he worked in Jefferson during much of the nesting season. The next year, 1853, Kumlien reported to Brewer: "I went over to the lake the other day on an egg excursion and found that the Xanthocephalus had not yet laid their eggs. I saw several but they did not seem to have any attachment to any certain place nor did I see a nest."[31] He knew that the birds would be noisily territorial if he had been near a nesting colony. He did, however, shoot one female. When he took it home to skin, Kumlien found that, "She had an egg in her about the size of a kernel of corn and four smaller ones." He was getting closer. The birds were breeding somewhere nearby. "If nothing hinders me," he wrote, "I will go there again in a week or two. I have hope of getting their eggs and learning more about their manners."[32] He knew he needed to spend time observing the behavior of these birds, not just collecting their eggs. But these birds nesting in the center of an extensive cattail marsh were difficult to see. He wrote Brewer, "The nest and eggs of the Xanthocephalus I could not find although I visited the place three different times, and I assure you I did my utmost searching for them."[33]

In June 1854, Kumlien had better luck. When he left his cabin in the woods above the spring with his collecting bag, shotgun, and maybe a dip net in hand, he walked down the trail to the lake shore where his rowboat was tied. Putting his gun and bag in the boat, he rowed to the marsh where he had seen the birds.

When he got to the edge of the marsh, he got out of the boat and stepped into the water, pushing the boat ahead of him. In water two to four feet deep, he pushed from the back of the boat so that it parted the tall, dense reeds ahead of him and kept his profile low, allowing him to get closer to the birds.

Near the center of the marsh, he found the yellow-headed birds nesting in colonies surrounded by colonies of red-winged blackbirds. In reeds so tall he could not see over them, Kumlien quietly pushed the boat ahead of him, wading in water over his knees. The birds did not seem to mind his presence, he wrote, so sometimes he got within ten feet of them. The males perched on the tops and stems of the cattails, displaying and defending their territory. He described the call of the male yellow-headed blackbird to Brewer, who had never heard it, as "something between the note of the redwing and a young rooster, the latter part of the notes being very hoarse. It sounds as if he were endeavoring to get out something nice. The commencement goes well enough but comes very near choking him before he gets through."[34]

"I have after several unsuccessful attempts to find the nests of the Xanthocephalus, now found it," Kumlien wrote Brewer in 1854.[35] Kumlien wrote that he had for Brewer a completed nest with four eggs, which he had found just before the middle of June when the "large young ones had left the nests."[36] Later that June, Kumlien found three more nests with eggs. The well-disguised nests Kumlien finally found were made of last year's floating cattail leaves woven between five or six upright cattails, making a nest five or six inches across that hung suspended over the water. The pale green or pale grey eggs had brownish mottling.

Taking care to keep the eggs and nests together, Kumlien carried them and the birds he shot home in his boat. When he got home, he labeled the nests and the eggs with the site and date. He measured the length of the birds from beak tip to tail. "The males," he reported to Brewer, "vary in length from 9½ to 10¾ in. The females are from 8½ to 8¾ at the most."[37]

Then, probably working outside on a table where the light was better and where he could use the arsenic outside of the house, he set out his small sharp knife, his scissors, and perhaps a pair of forceps or tweezers, and he set about skinning the birds. First, he laid them on their backs and began by making a careful and shallow cut through the skin of the belly from the vent to the ribs. Then he began to separate the skin of the upper legs from the muscle, being careful not to tear or stretch the skin. He was essentially turning the bird skin inside out by separating it from the body. He wanted to remove everything from the skin but the skull, wings, and legs, leaving enough bone to make sure the feathers of the tail and the wings did not fall out. He carefully removed the eyes and the brain, leaving the eyelids. This was delicate work requiring Kumlien's manual dexterity, knowledge of bird anatomy, patience, and aesthetic sense. And experience—he had been doing it since he was a boy.

As he worked, he used sawdust to absorb blood and fluids, which could discolor the skin. When the skin was cleaned of fat and any tissue that could rot, he washed the skin and dried it in the air or with more

Xanthocephalus
xanthocephalus BP.

♂

Förenta Staterna, Wisconsin,
maj 1864.   T. Kumlien.
13025.

This yellow-headed blackbird, held at the Swedish Museum of Natural History in Stockholm, was collected by Kumlien at Lake Koshkonong in May 1864. NRM 538973, SWEDISH MUSEUM OF NATURAL HISTORY moisture-absorbing sawdust. The dry bird skin was then completely dusted on the inside with powdered arsenic to preserve it from insect infestation. The bird was stuffed, ideally with cotton batting, but if Kumlien didn't have that, he could use "any soft, light, dry vegetable substance . . . rags, paper, crumbled leaves, fine dried grass, soft fibrous inner bark . . . or thistle down."[38] After the bird was stuffed with just the right amount of stuffing, and, if it was a larger bird, sewn shut, Kumlien smoothed the feathers, straightened the neck and body, arranged the wings and feathers close to the body, in some cases tied the legs (the left leg over the right), and either sewed or tied the beak shut. He tied a label to one of the legs.[39] The ornithologist Elliott Coues wrote that a skilled bird skinner could do about fifty birds a day, but not every day. It was demanding work, with the potential dangers of nicking yourself or inhaling arsenic. In the summer, Kumlien had no way to preserve a bird to be skinned at another time, so after a day in the field, he had to work in the evenings as long as the light lasted. In the fall and winter, he must have worked indoors by fire or candle light.

After the skins were finished, he may have wrapped the birds in a cylinder of paper, if he had paper. But it's likely he did not. Kumlien then packed the birds in wood boxes, which he probably made himself, using whatever packing material he could find. Lists of the boxes' contents were sent in letters and may have been included in the shipments to Boston and Washington, to Troy, New York, to Uppsala and Stockholm, Norway and Germany, and the Netherlands. Kumlien then had to use a wagon to take his boxes to the post office in Albion or Fort Atkinson or to a freight office. All of this was a huge amount of work. Figuring in all the time and materials, it's doubtful that Kumlien made much money. Yet he wrote to Brewer in 1854, "I am glad to get fifty cents apiece for yellow-headed blackbird skins and I wish I could sell many for that price. It is easier for me to kill a bird and skin it than it is to go out and work hard for 50 cents a day for a farmer."[40]

By August of 1854, Kumlien had sent Brewer three nests of the yellow-headed birds with their eggs: "in two of them there was one egg apiece, and in the third nest were four eggs."[41] Because of these nests with their eggs, and because Brewer trusted Kumlien's identifications, Brewer was able to correctly identify some misidentified eggs he had received from someone else. In the 1874 *Birds of North America*, Brewer mentions several sets of eggs and nests that had "come from Mr. Kumlien from the shores of Lake Koshkonong."[42]

In November 1854, it was noted in the proceedings of the Boston Society of Natural History that Mr. Thure Kumlien donated to the museum the skins of a Cooper's hawk, a pigeon hawk, a shore lark, a black-throated green warbler, a ruby-crowned wren, the eggs of the common American gallinule, the pied-billed grebe, the pinneated grouse, an American black tern—and the nest of a yellow-headed blackbird.[43]

10

BIRD PUZZLES

1853–1859

Confirming the identities of the North American birds was the work of early ornithologists such as Thomas Mayo Brewer, John Cassin, Spencer Fullerton Baird, and Thure Kumlien.

The study of birds in Europe had been carried on in large part within great institutions. Taxonomists compared vast collections of bird skins to type specimens at centers such as the natural history collection at Lund University in Sweden, the National Museum of Natural History in Holland, and the British Museum in London. Because the Smithsonian Institution and the Philadelphia Academy of Science had only recently begun to collect specimens of bird skins, eggs, and nests in the 1850s, North American ornithologists had learned about birds from their own observation, from studying the bird books available to them, and from meeting others with similar interests in natural history societies. For example, in Boston, William Brewster, an ornithologist with whom Kumlien later corresponded, first learned his birds by meeting with two friends once a week to read aloud and study John James Audubon's 1839 *Ornithological Biography* and Thomas Nuttall's 1832 *Manual of the Ornithology of the United States and of Canada*.[1] One could study plants at Harvard with Asa Gray, but there was no place in this country to go to school to study birds.

The oldest natural science institution in America was the Academy of Natural Sciences of Philadelphia, founded in 1812. By the late 1840s, it held a large collection of natural history specimens in a modern fireproof museum. Here, curator John Cassin identified and classified bird skins from around the world. The Boston Society of Natural History, founded in 1830, built its first museum in 1850. Brewer, a member of the Boston society since 1835, contributed many articles to its scholarly journal. The Smithsonian Institution, founded in 1846, was led by the first Smithsonian secretary, physicist Joseph Henry. Henry hired Baird, one of the most distinguished naturalists in the country, who dedicated the institution to original research. Baird, also the Smithsonian's first curator, was a serious collector, who arrived at the Smithsonian in 1850 with a natural history and antiquities collection that filled two railroad boxcars.

North American ornithologists—all at this point self-taught—were not only collecting bird skins, eggs, and nests from around the country but also were gathering "life histories" of birds. Life histories are bird biographies containing all available data on a bird species, including notes on physical description, reproduction, habitat, behavior, and distribution. To write these life histories, ornithologists relied on their own observations and those of men such as Kumlien around the country. Kumlien not only collected eggs and nests and skins but also collected what he saw and heard in the woods and the swamps, on the lakes and on the prairie. And he sent his observations in letters to Brewer. It is in letters like these that so many life histories were drafted. As Kumlien corresponded with Brewer, we can see him solving puzzles through the act of writing out what he knew, by reading and rereading Brewer's *Wilson's*, through hindsight as well as new observations, and through the exchange of information.

On March 15, 1853—nine years after the birth of their daughter Agusta, who died before turning two—Christine gave birth to their first son, Aaron Ludwig Kumlien. This son, called Ludwig, would become Kumlien's protégé, a talented ornithologist, and Kumlien's companion in observing and collecting the birds of Koshkonong.

Around this time, perhaps overwhelmed with being a new parent with little money in a crowded log cabin, Kumlien wrote to Brewer sounding almost Dickensian: "I wish you to understand my position. . . . I am poor Sir I have to work hard to support my family and I see money but seldom—I was not brought up to work."[2] Yet he was optimistic. With rental income from the new purchase of Bjorkander's forty acres north of Kumlien's original forty, with some additional work, and with what he got from selling bird skins and bird eggs to Brewer and others, he could support his family. They would live well, he wrote, "being content with little."[3] He wanted the income from this additional property and his winter work away from the farm to buy him "time for birds & flowers," of which he had been "passionately fond ever since a child." He added, "*quo semel est imbuta recens servabit odorum testa diu.*"[4] The Latin quotation is from the poet Horace and translates to "The first scent you pour in a jar lasts for years." That same scent was being poured into the jar that was little Ludwig.

In response to Kumlien's reports of his poverty, Brewer wrote in January 1854 to report that his brother owned a three-hundred-acre farm fifteen miles south of Madison. He asked if Kumlien would want to "undertake the management."[5] Kumlien thanked him for the offer in February but said he had only one yoke of oxen and that "a large farm requires more power than I have at present."[6]

But later that summer, Kumlien was able to think about birds and not about poverty. One day in August, he sat down to write a long letter to Brewer. He wanted to ask about the black tern.

Out in the cattail marsh or the wild rice marsh of Lake Koshkonong, Kumlien had seen the black tern—one of the few Wisconsin birds that he knew from Sweden.[7] He recognized the almost dragonfly-like flight of the graceful little tern, which has been described as the "restless waif of the air."[8] While chasing a moth, it will dart and zigzag like a flycatcher. Over the water, it skims like a swallow. And like terns hovering over ocean waves, it can hover over a sea of wild rice, "dipping down occasionally to snatch an insect from the slender, swaying tops."[9] Kumlien wrote Brewer that he had skins of the black tern.

At least, he *thought* it was the black tern, but there were differences that puzzled him. He wrote: "I have myself shot St. nigra [black tern] in three different parts of Sweden and this seems to be the same as far as I remember."[10] He must have been referring to Lake Hornborga, Gotland Island, and some third place. Kumlien had skins for Brewer, and some questions: "In your synopsis in Wilson you say head, neck, breast, sides and abdomen *grayish black* length 9 inches."[11] In his copy of Nilsson's *Scandinavian Fauna*, Kumlien read a quite similar description of the black tern: "bill black, head and neck black lower parts from the bill to the lower tail coverts blackish gray, upper parts from the bill to the lower tail coverts blackish

This black tern was collected by Kumlien on June 6, 1862, almost exactly twenty years after he collected black terns on Gotland Island. NRM 536322, SWEDISH MUSEUM OF NATURAL HISTORY

gray, upper parts back wings and tail ash colored or lead color. Length 9½."[12] But the tern skins Kumlien collected at Koshkonong were lighter colored and larger: "I have specimens here with the head neck breast and abdomen black, underwing coverts very light ash and the outer edge of the wings from the body to where the primaries commence nearly white and I have specimens varying in size from 9¾ female to 10¼ to 10½ male. The eggs too vary considerably, some being a deal darker than others."[13]

Kumlien was then interrupted in his letter writing: "Just now came a boy bringing one dozen quail [northern bobwhite] eggs he found while making hay this forenoon." The interruption reminded Kumlien that he had eggs of the tern he was puzzled about. He continued to Brewer: "Would you like about 15 to 20 eggs of the Sterna [black tern]?"[14] Kumlien had answered his own question—he was sure that what he had were color variations of both the skins and eggs of the black tern.

Kumlien was open with Brewer, admitting that he was also collecting for Henry Bryant in Boston—"if it does not interfere with my engagements with you."[15] A month later, Brewer wrote that Bryant "is rich and can well afford to pay you handsomely, but is uncertain and erratic." Brewer added that he would take two-thirds of the black tern eggs Kumlien had collected but told Kumlien to send the rest to Bryant, "who will pay better."[16]

At this time, Kumlien was also corresponding with several other American collectors. One of them was George B. Warren, president of the Troy

Union Railroad Company in Troy, New York, who wanted eggs and skins for his private collection. The critical and imperious Warren also collected miniature antique Chinese pottery. In August 1854, Kumlien told Brewer that he had sent Warren the skins and eggs of thirty or so common birds, as well as the skin and eggs of a bird a neighbor had found and given to him. Kumlien wrote that he sent the skin and eggs to Warren thinking them "of little or no consequence having seen three or four such nests before though I could not find the bird in the book [*Wilson's American Ornithology*] which I attributed to the poor state of the skin and my ignorance of the Sylvia."[17] Though Kumlien had identified the skin and eggs he sent to Warren as those of the yellow warbler, he knew the eggs were different. For two years, he read the list of warblers over and over in his copy of *Wilson's*, finally coming to believe that "the bird I let go with eggs and nest was nothing less than S. Rathbonii"—*Sylvia Rathbonii*, or Rathbone's warbler. But he told Brewer not to worry: "I have one egg left and if you desire it you can have it."[18] Brewer responded that he was quite curious about the *Sylvia Rathbonii* and hoped Kumlien could get another. He then added that neither Cassin nor Baird "believe in any such bird and it did not become me to differ." Brewer, who knew of Warren, wrote that it was "a pity" the bird in question had "fallen into hands so little able to make it available"[19]—that is, into Warren's private collection, which was not available to scientists.

The fact that the eggs Kumlien found differed from his book's description of yellow warbler eggs steered him away from his original, and very probably correct, identification of the yellow warbler. But Cassin and Baird turned out to be right: there was no such bird as Rathbone's warbler. Audubon had first named and painted it at the Falls of the Ohio in 1808. But what he had painted was two juvenile yellow warblers he had mistaken for a new species and named Rathbone's Wood-Warbler.[20] Another broken link in the chain of Kumlien's identification of the bird is that the specimen Kumlien sent to Warren with eggs and nest had been gathered by a neighbor—a neighbor who perhaps gathered a nest here, some eggs there, and a damaged yellow warbler specimen somewhere else. Kumlien acknowledged to himself and to Brewer that in the case of this warbler, the chain of steps that might have led to its correct identification had broken down. And though this had happened two years earlier, Kumlien remembered his mistakes and regretted that he had failed to correctly

When Audubon painted these juvenile yellow warblers at the Falls of the Ohio in 1808, he mistakenly believed he had discovered a new species, which he named Rathbone's Warbler.
NATIONAL AUDUBON SOCIETY

identify the birds and eggs, which were lost to science within Warren's decorative collection.

Though Warren may have been simply collecting lovely objects when he acquired bird eggs, Brewer, Baird, Cassin, and the other ornithologists of the time were not only collecting the physical birds eggs and skins but also collecting observations on breeding locations and times, migration, and all aspects of the birds of the newly settled areas of the country. Many observations Kumlien sent to Brewer would be later used in Baird, Brewer, and Ridgway's *Birds of North America*, which would contain extensive life histories of the birds written by Brewer. Knowledge of breeding plumage, which differed in many birds from fall and winter plumage, and juvenile plumage at this early stage in the study of birds in America was often not detailed

Despite his initial confusion over the yellow warbler, Kumlien did successfully identify the Philadelphia vireo and distinguish it from other species. The story begins in 1842 in a place called Bingham's Woods outside of Philadelphia. Cassin, then twenty-nine years old, was out looking for migrating fall vireos, a family of little yellowish-green birds, hard to tell one species from another. Cassin spied a a family of little gray and yellow bird high in a tree, moving slowly and deliberately rather than in the busy darting manner of so many other little birds. He shot and skinned the bird and held it in his private collection until 1851 when, in a short article for the Academy of Natural Sciences, he described the bird as a new species he named *Vireosylvia philadelphicus*. Cassin never saw another Philadelphia vireo other than the one he shot. He believed it to be a rare bird.

Yet it was not a rare bird then, and it is not a rare bird now. The Philadelphia vireo, as it is now known, has a wide geographic range—breeding all across southern Canada, migrating through much of the eastern half of the United States, and wintering in southern Central America. The bird's secretive habits made it seem rare to early ornithologists who hadn't seen any or many.

Cassin first brought this bird to light in 1851 when he described it in a short paper for the *Proceedings of the Academy of Natural Sciences* in Philadelphia.[21] Kumlien did not see this publication, nor had he read any description of the bird because it was not described in the only work he had on American birds, his 1841 edition of Brewer's *Wilson's Ornithology*. Kumlien sent specimens of the Philadelphia vireo to Brewer in 1854 and 1855, but neither Brewer nor Kumlien knew what they were. Brewer identified three of the specimens as *Vireo gilvus*, or the warbling vireo, "in unusually fresh plumage."[22]

Kumlien knew that it was not *gilvus* and argued this point in a letter to Brewer.[23] But Brewer, who was only looking at skins, was not convinced. Then Kumlien, who had observed both *gilvus* and *philadelphicus* in the field, wrote his complete reasoning for the differentiation of the two species to Brewer. Kumlien was so convincing that not only did Brewer believe the birds to be the *philadelphicus*, but he also entered Kumlien's well-written letter into the *Proceedings of the Boston Natural History Society*.

In the 1859 *Proceedings*, Brewer set the background. Two years earlier, Kumlien, "a very accurate and careful ornithologist of Wisconsin," had sent Brewer a specimen of vireo he had obtained near Lake Koshkonong, a species Kumlien had never seen described, one "quite distinct from the *V. gilva* [warbling vireo]." Brewer then told how he gave the specimen to a friend whose opinion he valued highly, and this friend said it was *V. gilva*. When Brewer conveyed this information to Kumlien, Kumlien still "insisted that its habits, even more than plumage and size, showed it to be a distinct species." The following year, Brewer gave Cassin several new specimens Kumlien had sent. Cassin "had no doubt they were of the species he had described as *V. Philadelphica* [Philadelphia vireo], though others . . . were still unconvinced." Brewer continued, "In answer to a letter in which

Kumlien collected this Philadelphia vireo (left) on September 19, 1863, and this warbling vireo on May 15, 1854. He identified these vireos in letters to Thomas Mayo Brewer. NRM 534422 AND NRM 534382, SWEDISH MUSEUM OF NATURAL HISTORY

I informed Mr. Kumlien that his birds were supposed to be *V. gilva* in an unusually fresh plumage, he wrote me the answer which I give below."[24]

As Brewer knew, Kumlien's letter on the Philadelphia vireo was a fine demonstration of clear scientific reasoning. Kumlien first demonstrated that he knew the *Vireo gilvus* very well. Kumlien told Brewer that the *gilvus* "in every respect" agreed with Wilson's description of it. He said it was common at Koshkonong from May 8 or 10 until September. Those dates told him that "it consequently breeds here." Kumlien had heard its song: "it is an excellent singer." He also had a number of its skins: they were alike in their markings, and they agreed with Wilson's description of *gilvus*. Kumlien had *gilvus* skins from spring and fall and knew that there was very little difference between spring and fall plumage. He had found the bird in "openings more than in thick timber and frequently near farmhouses." He had measured the length of the skins and found that the length "varies from 5½ to 6 inches." He said he had one that "measures full six."[25]

The second unnamed vireo, Kumlien argued, "is by no means so common" as *gilvus*. Kumlien had "never observed it before May 15th, and only from the 15th to the 25th of May," and then in September. These dates told him that this bird was not breeding at Koshkonong. When Kumlien saw it, it was migrating. He said he had never heard this bird sing a note. He had found it nowhere but "in the most secluded thickets." Additionally, this bird was smaller than the *gilvus*, with a length "from 5 to 5¼ inches, which is the longest I have ever found."[26]

Kumlien conceded that the markings of both birds were very similar, "but the *gilvus* is a more slender bird than the other, which appears stouter." This comparison of body types is still used to distinguish the two birds. Also, when he saw the unnamed bird in the autumn, it was "considerably tinged with yellow." Kumlien then added a chart listing the differing lengths of the five primary feathers and the wing on each bird. "This will, I think, separate them," he wrote. "This measurement was taken from several specimens."[27]

Kumlien made one more point to demonstrate his ability to distinguish one species from another: "But the question may arise, is not my *Vireo gilvus* not *V. noveboracensis* [the white-eyed vireo, *Vireo griseus* in modern usage]?" This question had arisen when Brewer asked it in a letter to Kumlien in 1854: "The matter of the *vireo gilvus* is worth studying another season. Is not the other species you speak of the white-eyed vireo?"[28]

"It is not," wrote Kumlien in his response. "The iris is hazel and not white, and moreover, on my *gilvus* there are no yellow markings, except a very faint greenish-yellow on the breast. *V. noveboracensis* is 5¼ inches in length, and I have never had a specimen of *V. gilvus* as small as that."[29] He didn't bother to say that the unnamed vireo was not the white-eyed vireo.

In his article for the Boston society's *Proceedings*, Brewer concluded, "I have given Mr. Kumlien's letter in nearly his own language, and in no instance have I varied from his meaning. I think it establishes his vireo to be a good species, and if so, it is the *Vireo Philadelphica* of Mr. Cassin."[30]

Brewer ended by writing, "I take the greatest pleasure in thus giving Mr. Kumlien the credit of having worked it out, unaided by any suggestion or help from any one, in view of the disadvantages under which he labors in the want of access to any text-books. His letter throws 'the first light that has yet been given to the public upon . . . this new and little known species.' "[31] Kumlien's was the first description of the "habits and distribution" of the Philadelphia vireo.

Though the Philadelphia vireo is now known to be abundant and to have a wide geographic range, it "remains one of North America's most obscure birds."[32] Long after Kumlien was gone, it would take years of bird banding before ornithologists worked out that the Philadelphia vireo breeds in Canada, farther north than any other vireo, and it winters thousands of miles south in southern Central America. Both Kumlien and Cassin had seen the Philadelphia vireo on its journey of thousands of miles.

The 1850s were a busy decade in Kumlien's life. Though most of his correspondence during this time was with Brewer,[33] Kumlien also wrote and sent specimens to collectors Warren in Troy, New York, and Bryant in Boston. Baird at the Smithsonian also exchanged letters with Kumlien in 1855: Baird wanted a shrew specimen, and Kumlien wanted books on natural history.[34]

Meanwhile, during all of this observing, collecting, preserving, and shipping of bird specimens, Kumlien's family was growing. On January 7, 1855, Thure and Christine's second son, Theodore Victor Kumlien, was born. And a daughter, Swea Maria Kumlien, was born seventeen months later on August 8, 1857. A little more than two years later, the Kumliens' third son, Axel Frithiof Kumlien, called Frithiof, was born on December 19, 1859.

But before Frithiof's birth, in early spring of 1859, Kumlien received a letter from Carl Löwenhielm that must have provided him some relief. In the letter, Löwenhielm acknowledged several shipments of bird specimens Kumlien had recently sent. Included was the loan note signed by Kumlien on May 29, 1843, acknowledging his debt to Löwenhielm for paying Kumlien's passage to America. "At the bottom of the note," Kumlien's granddaughter observed years later, "is a notation by Lowenhielm on February 11, 1859, stating the money had all been paid by Thure Kumlien by his sending collections of birds and other objects of natural history."[35] After sixteen years, Kumlien's debt for passage to America had been paid.

11

FRIENDSHIPS, BOTANY, AND WAR

1859–1866

In his 1850s correspondence with Thomas Brewer, we see Thure Kumlien building on his abilities as a field ornithologist and finding a confident voice as a scientist. He was lucky that it was Brewer who first sought out his birds and observations and not Bryant or Cassin or one of the other colder and more critical eastern big shots. Brewer's warmth and generosity, as well as his professionalism, elicited not only bird eggs and bird skins from Kumlien but also his best ornithological observations and his fluency in written English, particularly in his letters about the Philadelphia vireo.

Though Kumlien continued to correspond with and collect for Brewer and a few other Eastern ornithologists, and though he sent large shipments of specimens to European museums, the most interesting and revealing aspect of Kumlien's life in the 1860s is his correspondence with two botanists—Elias Fries, his friend and former botany professor at Uppsala University in Sweden, and Edward Lee Greene, his young botany protégé and neighbor. Each series of letters began in the 1860s—one in spite of the Civil War and the other because of it. Both Fries and Greene were botanists rather than ornithologists. Their interests in plants and their requests of Kumlien swayed him to focus on plants rather than birds over the years of the war. And his communication with these two correspondents reveals Kumlien in ways that his other correspondence does not. In these letters, we see Kumlien as both teacher and student, the central figure of three generations of botanists—Fries in Sweden nearing the end of his career,

Kumlien in Wisconsin nearing the height of his career, and Greene setting out from Wisconsin to the West at the beginning of his great career. And we see Kumlien rediscovering botany, his other first love.

In the Fries correspondence, we have only Kumlien's letters; in the correspondence with Greene, we have only Greene's.

In the 1860s, the world was widened for Kumlien and for other settlers. Travel was easier. Roads and bridges had been built or improved. And railroads connected Milwaukee to Milton to Janesville, connected Milton to Stoughton to Madison to Prairie du Chien, and connected all of them to parts east and west. Wheat crops were easier to ship and sell. Mail was faster and more reliable. And there were more people in the country. The population of Jefferson County had doubled to more than thirty thousand since 1850. An increasing number of newspapers were read by this increasing number of people. Besides the *Janesville Gazette*, Kumlien could read the *Weekly Jeffersonian*, the *Wisconsin Chief*, and the *Fort Atkinson Standard*. Most settlers had replaced log cabins with frame houses built on improved roads. But the Kumliens still lived in their log cabin on the Indian trail in the woods.

A log schoolhouse had been built in 1855, not a mile from the Kumlien cabin, which seven-year-old Ludwig attended in 1860, soon to be followed by his brother Theodore, sister Swea, and brother Frithiof. Kumlien's post office address in 1860 was Albion in Dane County, a town located about four miles west of Kumlien's cabin.

A group of well-educated New Englanders had settled in and around Albion, establishing an academy in 1854. Among them was the Green family, William and Abby and their children, who had come from Rhode Island in 1858. The Kumliens and the Greens became friends, but the closest friendship among them was that between the Greens' fifteen-year-old plant-mad boy, Edward, and the forty-year-old Kumlien. Edward Lee Green—who would later change the spelling of his name to Greene—was born August 20, 1843, the day the *Svea* arrived in New York. He and Kumlien became botanical companions, Kumlien teaching young Edward as they sought out the native plants of the rapidly shrinking prairie and savanna, the hardwood timber, the damp meadows and the tamarack swamp, and on the shores and in the waters of the chain of three small lakes between their homes. Joining them on some of their excursions was

Kumlien's son Ludwig, who was described by his father as having "a complete mania for flowers and all sorts of natural history specimens."[1]

In early August 1859, Kumlien was surprised by a letter from Fries, who was now the director of Hortus Botanicus, the Uppsala University Botanical Garden founded by Carl Linnaeus.[2] Sixteen years after Kumlien had arrived in Wisconsin, Fries wrote to ask his former student to collect pressed and live plants for the famous botanical garden in Uppsala. "I will do it as well as I possibly can," Kumlien responded. And he asked that "Mr. Professor excuse my writing. It is so seldom that I write in Swedish that I fear that I have made big mistakes."[3] Eight years earlier, Kumlien had asked Brewer to excuse his seldom-used English. "Much have I of course forgotten," he wrote of his botany, "but the desire remains and if I were to collect plants again and learn botany thoroughly it would give me great pleasure."[4] He said he would put together a small collection for Fries that fall.

Kumlien closed his first letter to Fries: "With greatest respect I have the honor to remain the Renowned Mr. Professor and Knight's grateful servant, Thure Kumlien." The formality of this language gives us a sense of what Kumlien believed his status to be relative to Fries's. He did not assume an American voice but reverted to the voice of a Swedish student writing to a famous high-status professor. And this tone remained consistent in the four letters of Kumlien's that we have, which stretch over the next few years.

In order to collect a good representation of the plants and seeds of various habitats, Kumlien knew that each would have to be collected several times a year. He did not hesitate to commit himself to the time, expense, and work involved in adding collecting plants to his tasks of collecting bird eggs and bird skins, farming, and raising four children. He would not turn down his old professor. Fries not only was one of the most important botanists in Europe, but he was also an old friend. Kumlien, too, loved plants. And he needed the money—though he didn't ask Fries what that money might be.

Fries had apparently sent a catalogue or wish list of plants for the garden and museum. In response to this list, Kumlien wrote Fries that the "*Saracenia* [pitcher plant] is to be found here, but when I the other day visited the bog where it grows I couldn't find any stalk with seeds."[5] He would look for it in a place where the cattle hadn't grazed. But he knew where to find water shield, yellow lotus, and some orchids (white adder's tongue and snakemouth) in bogs and tuberous grass pink in damp meadows.

Fries also wanted specimens of *Cistaceae*, but Kumlien had seen only one or two of these sand dune plants. He tactfully explained to Fries that he lived in what was called "openings land," or savanna, where the trees (oak and hickory) had an understory of hazel. Nearby were the prairie and the tamarack bogs. To the northeast—thirty or so miles—was "thick timber."[6] Even farther north was pine and spruce forest.

"Here we have a large amount of . . . sedges," Kumlien wrote, "but I have not examined any of them."[7] He knew a sedge when he saw it, but he had not taken the time to distinguish the many difficult species. If he had not studied North American plants with the intensity that he had studied the birds, still he knew *how* to study them, and he had some references: "I own Wood's *Class-Book of Botany*."[8] Also, Brewer had sent him a used copy of Asa Gray's important and useful *Botany of the Northern United States*, usually called *Gray's Manual*. Kumlien had two books by Fries himself, books he had brought with him from Sweden—*Flora Scanica*, on the plants of Sweden, and Fries's great work on the lichens of Europe, *Lichenographia europaea reformata*, with his groundbreaking system for classifying lichens based on the characters of their fruiting bodies.[9]

Kumlien's greatest resources were his powers of observation and his memory for not only what he had seen but also where and when he had seen it. He had been tramping over the country around him for nearly twenty years, so he knew what grew where as well as what *used to* grow where. He wrote, "The country around here, for miles, has long been bought and if it is not turned into farm fields it is grazed with the result that many plants have disappeared which used to grow here in abundance."[10] Plowing and grazing had destroyed the wild lupine, the showy lady's slipper, and the Indian pipe, which Kumlien had found easily in the 1840s but by 1860 was forced to search for in fence rows and in the corners of fields on patches of ground where agriculture had not reached.

Kumlien told Fries that one of the reasons he wanted to collect for Uppsala was that the income might enable him to travel and collect in the western states, the Mississippi Valley, and the south. A hard-core naturalist, Kumlien wanted to turn income into time in new and unexplored fields. This desire to see new country comes up repeatedly in his letters of the next few years. Perhaps Kumlien's often stated but seldom realized longing to travel to new places reflects his real desire to find and collect species new to him that he couldn't find close to home. But he also may have consciously or unconsciously wanted to travel to places where nature was intact, untouched by settlement and European agriculture, where he didn't have to feel the constant little griefs of remembering what had once been but now was gone.

And Kumlien had responsibilities from which he might have wanted a temporary escape. Though he had enthusiasm and the best of intentions, it wasn't until March of the following year that Kumlien got his first collection off to Fries. Kumlien explained that chores on his small farm took all his time. And by the time he was ready to ship, a long stretch of illnesses in his family had "made it impossible to even think about natural history."[11]

The box that he sent to Fries on March 6 contained "94 types of seeds, about 130 dried phanerogams [seed-producing plants], some mushrooms, about 30 lichens and some mosses, and also a can, containing many live plants of *Sarracenia purpurea* [purple pitcher plant]."[12] He also sent a collection of birds for Carl Gustaf Löwenhielm, who was in Uppsala at that time. Kumlien sent Fries a numbered list of the lichens, mosses, and mushrooms included in the box with a duplicate list on which he asked Fries to fill in any names Kumlien didn't know and send the completed list back to him. Kumlien instructed Fries to plant the live pitcher plant included in this shipment in a swamp or it would die as had one he'd tried to transplant into his own garden. Though it is rarely mentioned in his writings, Kumlien had a garden of ornamental plants that he wanted to keep an aesthetic or botanical eye on.

In a March 27 letter to Fries, after Kumlien described the contents of the box he had shipped on March 6, he looked ahead to the coming collecting season. Though the ground was still frozen and covered with snow, Kumlien said he would immediately begin collecting "as soon as any plant will be ready, which perhaps will happen in a few days." From years of observation, Kumlien knew the order of the flowering. First would come the willows, he said, and the hazels. Then the maples and other trees. Then the pasqueflower, the prairie buttercup, and the sedges. Kumlien wrote that he would do anything he could to make the plants he sent his professor

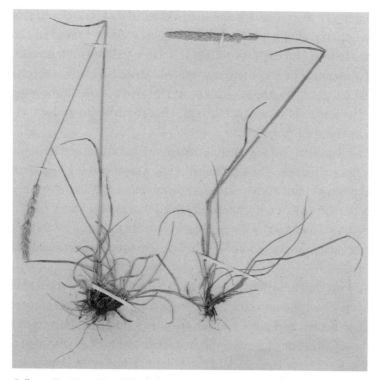

Collected by Thure Kumlien near Albion in Dane County, this native June grass is one of the plants Kumlien sent to Elias Fries and the Uppsala University Herbarium in the 1860s. Nineteen sheets of these plants are held at the Swedish Museum of Natural History. HERBARIUM SPECIMEN S13-20012, IMAGE BY HEATHER WOOD, SWEDISH MUSEUM OF NATURAL HISTORY

"beautiful, complete, and well pressed."[13] He would collect for him as many seeds as he could find before October when he planned to ship the summer's collection.

As Kumlien was writing this letter, his eight-year-old son Ludwig came into the room with a blooming pasqueflower: "the first this year even though the ground is frozen almost everywhere and we had snow last night."[14]

Kumlien then told Fries about a subtle environmental effect he had noticed since the Native people were no longer burning the prairie and savanna:

It is peculiar that in the last 8 to 10 years the lichens around here have increased! The trees show many yellow *Parmeliae* now which were not to be found before. It may have to do with the so called "Prairie" fires nowadays being over. In the old days they used to sweep over the land every year and burn everything. Since the land has been bought and fenced these fires have ceased. It is only a few years ago since I observed the first *Peltigera* but now there are at least 2 kinds, both rather common.[15]

It is likely that no one else had documented this effect of white settlement on lichens on savanna trees, reminding us that Kumlien saw this land at the end of an era, the way it had been for thousands of years.

In the late spring or early summer of 1860, Kumlien got a large order from the *Naturhistoriska riksmuseet*, or Swedish Museum of Natural History, in Stockholm to collect zoological specimens.[16] With the promise of this money from the museum, Kumlien was able to arrange to be free from all farm work that spring and summer: "I have hired a man to do this for me," he told Fries.[17]

After this order, Kumlien made two trips in June, "about 30 English miles northeast from my home," into what he called the "thick forest." He wrote Fries that he found a lot, "but what I also found was that I should have been there 2 or 3 weeks earlier."[18] The thick canopy of a mature forest dominated by sugar maple and American beech allows for the growth of few understory plants except the spring ephemerals, which bloom before the canopy leafs out. Kumlien had missed the magnificent bloom of the spring ephemerals in a northern hardwood forest.

For the rest of the summer and early fall of 1860, Kumlien visited nearby and well-known habitats collecting plants and birds and small animals. In the remnants of bogs and swamps, in meadows too wet for plowing, on bluffs too steep for grazing cattle, at the edges of the little lakes, in fence rows, on the high prairie, and on dry sandy hills, Kumlien sought the rich and disappearing life he had known since 1843. Though he had shot and trapped and cut and collected, his depredation was tiny compared to that of agriculture, and his collecting was in the interest of scientific knowledge.

On February 28, 1861, Kumlien began a letter to Fries saying he was glad the seeds had arrived and he hoped they would germinate, but after two paragraphs, Kumlien began his report on the previous summer's collection—plants for Fries and birds and animals for the Museum of Natural History in Stockholm. Carefully listing the almost five hundred specimens he sent to Sweden in 1860 must have taken Kumlien a whole winter of work. He kept one of each specimen, giving them individual identifying numbers, and in a separate catalogue he noted the exact locations where he had found less common species.

Even with time off from farming, Kumlien wrote Fries, he hadn't had time to carefully identify all the plants, particularly the difficult grasses and sedges. If he succeeded in finding anything really rare, he wouldn't know it for sure.

One thing bothered him:

> The Orchid, no 223, which I sent previously . . . and which I thought was *Microstylis aphioglossoides*, I now after more careful examination consider to be something else, which my book cannot tell me. I found it three years ago in July in a grazing meadow, but since then I have not found it again. No 331 grows in ditches and has a yellow flower and No 316 grows along lake shores in muddy locations and No 332 which I couldn't find in bloom; these are causing me trouble and I cannot find their names.[19]

When he wrote about the plants and their identification, Kumlien's language was direct, and so was the relationship he assumed with Fries— botanist to botanist. But sometimes Kumlien provided disclaimers, as when he said Fries could not expect that all the plant names were correct because Kumlien was out of practice and he didn't have books as good as the ones he had had in Sweden.

And sometimes Kumlien was apologetic, as when he asked Fries for his pardon and admitted that "many specimens are not in good shape entirely, but perhaps they are good enough to allow you to key them out to a species."[20]

Kumlien hoped they were good enough for Fries to sell. How many of each could he sell? Were the plants in the previous shipment in saleable condition? How long would he have to wait for payment when he sent plants in the fall to Uppsala?

Kumlien sometimes belabored a point with Fries that he previously belabored with Brewer: in order to do the collecting, he needed to hire help for his small farm, which was all that he and his wife and four children had to live on. When he was collecting, his budget was small and costs were high. An ordinary farmhand, for instance, cost twelve to fifteen dollars a month plus food, and some hired help couldn't even be had for that amount. "It is therefore important for me to know within at least 6 months," he wrote, "how soon after the arrival of the plants to Europe, I will be able to expect any payment, so that I can adjust to this from the start. I most strongly wish and hope that the R. Mr. Professor completely understands my position. . . . I wish that the suggested plant commerce works out—at the same time as it would help me support my family, it would be the occupation that I most simply wish for."[21] The real desperation in some of his letters to Fries must have made Kumlien want to call them back.

The Kumliens didn't have much money, and they lived in a small log cabin, somehow making space for collecting plants and small mammals, as well as birds, while raising four young children. It must have been at this time that Kumlien built a separate shed for his scientific work. We know he had such a thing in the 1870s, though we have no image of it. Everything he collected had to be cleaned, prepared, and preserved by drying or treated with arsenic or pickled in alcohol. Everything had to be identified and labeled, listed at least twice, and then stored until the whole season's work could be packed and shipped. Collecting seeds, too, involved careful drying, labeling, and storage. Plants had to be first arranged carefully on drying sheets, pressed in a plant press (two boards weighted or strapped together tightly), checked often for mildew and insect damage, then attached to sheets of paper with needle and thread or glued slips of paper and carefully labeled with date and location on separate sheets.

Kumlien collected insects as well. But without insect pins, they were difficult to ship and present undamaged. And insect pins—not just any pins—were expensive and hard to get in southern Wisconsin in the 1860s. When Kumlien began collecting small mammals, he used another set of supplies, equipment, and skills. Fish and amphibians he preserved and sent in jars of alcohol.

By the 1870s, Kumlien could order his birds' eyes and insect pins from this taxidermy supply house in Chicago. WHI IMAGE ID 147775

That summer of 1860, Kumlien collected and sent to Fries more than one hundred specimens of *Draba caroliniara* [Whitlow grass], *Houstonia caerulea* [azure bluet or Quaker ladies], *Mitella nuda* [naked miterwort], and other plants, but he had not found one specimen of *Synthyris* in the plantain family. They, like some orchids, had all been eaten by grazing cattle. Eventually, though, in October, on a steep hill close to a river, he did find "a whole heap" of *Synthyris*, "undisturbed and dry, only the large root leaves were fresh."[22]

In this letter and others in the early 1860s, Kumlien mentioned that he was expecting to find new plant species in Wisconsin when he traveled north toward Lake Superior or west toward the Mississippi River. He wanted to write a *Plants of Wisconsin*. When he told Fries in 1861 that there were no *Sabbatia* (probably marsh plants of the gentian family) growing nearby, he added that the *Sabbatia* "were not mentioned in a source I have about the plants of Wisconsin, which I know is not complete, because I have found several other plants that are not mentioned in it."[23] His source was likely the first widely available listing of the birds, plants, and animals of Wisconsin: Increase A. Lapham's "Flora of Wisconsin," published with Philo R. Hoy's "Birds of Wisconsin" in *Transactions of the Wisconsin State Agricultural Society* in 1852. Kumlien must have proposed sending the Boston Natural History Society a list of the plants of Wisconsin because Brewer responded in 1862 that "our society would be glad to have a list of the plants of Wisconsin whenever you can prepare it."[24] Nothing came of this plan, however, and it's too bad because Kumlien's unique knowledge of the original plants of at least the Koshkonong area can never be duplicated.

While he was working at collecting, Kumlien was also following in the newspapers the events leading up to the Civil War. He wrote Fries: "Here in America the Politics are bad. 6 of the slave states have left the Union and several others are believed to soon follow the same way. The new President who is about to take office on the 4th of March [1861] was close to being murdered the other day! Yet this turbulence will not harm us here in the Northwest very much!"[25] Though the "turbulence" would affect everyone in "the Northwest," all through the 1860s Kumlien would be able to go on collecting and sending boxes to Sweden, as if there were no war at all.

Soon after the surrender of Fort Sumter to the Confederates on April 12, 1861, on the same day President Abraham Lincoln sent a telegram to Governor Alexander W. Randall requesting Wisconsin volunteers to serve for three months, the *Janesville Gazette* and most newspapers in Wisconsin agreed that the rebels must be put down. Though this sentiment was common, there were racists in Wisconsin who feared that political equality for freed black people would mean their social equality with whites, some of whom wanted nothing to do with black people. But there was a stop on the Underground Railroad in Milton, just south of Lake Koshkonong, and black people who had escaped enslavement were also being aided by citizens of Janesville and Beloit. And at nearby Albion Academy, after driving out two students who were Confederate sympathizers, some faculty and students formed a secret society ready to aid people escaping from slavery. A rally of Janesville citizens in early 1861 drove away a "slave catcher" who had tracked down one of Janesville's residents who had been formerly enslaved. In Wisconsin, Minnesota, and Illinois, Scandinavian immigrants responded quickly and enthusiastically to Lincoln's call for volunteers.

Because so many men volunteered, there were labor shortages, and the cost of labor soared. Kumlien complained to Fries that "wages are *ravenously* high here. $40 per month for a simple worker during the harvest." He also reminded Fries, "The name of this state is Wisconsin, not Viscosin." Kumlien's voice in his letters to Fries was sometimes that of an American informing a citizen of the Old World, but at his most clear and direct, he was a fellow botanist trying to learn the name of orchid No 223: "I cannot find it anymore!—the roots were small tubes, the size of peas."[26]

Sometime in June or July of 1861, Kumlien received a long, discouraging letter from his friend Löwenhielm. Löwenhielm confirmed that a package from Kumlien had safely arrived at the Museum of Natural History in Stockholm and that a box for Fries arrived in Uppsala as well. Fries, Löwenhielm wrote, was pleased with its contents, and Löwenhielm thanked Kumlien for his share of the specimens in those boxes.

But he told Kumlien that Professor Carl Jakob Sundevall[27] at the Natural History Museum didn't want any more birds or animals from Kumlien because he didn't want more duplicates; instead, Sundevall might buy birds from a French dealer in natural history specimens, Auguste Sallé.[28] Also, there was another collector in Racine, Wisconsin, who had sent Sundevall a bird.[29] Löwenhielm recommended that Kumlien give up the collecting of birds and animals for Sundevall, but the next year and for several years after that, Sundevall ordered more bird skins from Kumlien.

Then Löwenhielm said another odd thing to Kumlien. He said he would mark Kumlien's debt from 1843 paid if Kumlien sent a list of all the specimens he had sent to Löwenhielm over the years. But Löwenhielm had already marked Kumlien's debt paid in 1859. Kumlien may have wondered about the state of his friend's recordkeeping or the state of his health. And in fact, Löwenhielm's health was not good. He wrote that he would soon go to the west coast of Sweden to "take salt baths." A doctor had suggested his illness may have come from "many years of breathing arsenic . . . living together with the birds."[30] Like most ornithologists of the time, and like Kumlien, Löwenhielm preserved bird skins using powdered arsenic, which can cause abdominal problems, diarrhea, heart disease, numbing of the hands, and cancer after long-term exposure. In Kumlien's communications over the years, he never mentioned any concerns about the effects of arsenic use on him or his family.

Though Kumlien had hoped to embark on "hunting excursions" to places as far away as Texas, in June 1861, he did travel about sixty miles northwest to the Baraboo Hills in Sauk County where he could "search among the high bluffs for preglacier flowers."[31] In that quartzite landscape grew plants unlike those around Koshkonong. Two plants we know he collected in Sauk County were rock harlequin and Gray's sedge.[32] Kumlien may have taken Greene with him on one of these trips.[33]

As he collected for Fries and Sundevall, Kumlien was still collecting for Brewer in Boston. On August 11, 1861, he told Brewer he had for him this season the common ground finch [Eastern towhee], the yellow-throated vireo, and the orchard oriole with two eggs and its nest "very neatly built of june grass." "If you don't care about them I will keep them," he added.[34] He told Brewer he was collecting insects for entomologist Carl Henrik Boheman at the Royal Museum in Stockholm. But he didn't have any way to get insect pins. Could Brewer help him with that? He also wrote that somebody sent him an advertisement for Baird, Cassin, and Lawrence's "splendid work." This well-illustrated 1860 publication consisted essentially of the bird sections from the Pacific Railway Survey Reports. The twelve volumes of the Pacific Railroad Survey would combine explorations and surveys to ascertain the best rail route from the Mississippi River to the Pacific Ocean.[35] Would any of the "gentlemen Editors," Kumlien wondered, be willing to trade birds or animals for a copy of the book? "I am willing to collect a good deal in order to get a good book," he said.[36] He also wrote Cassin offering to exchange bird skins for a copy of this work with colored illustrations, "as I have never seen any *plates* of N. American birds."[37] And he asked Cassin about insect pins.

In the 1861 box he sent to the Uppsala museum, Kumlien sent a small collection for Fries himself, plants collected in the Baraboo Hills. "I imagine that some of these plants are rare," he wrote. "I'm particularly interested to hear what the names of No 17 and No 62 will be."[38] Unfortunately, the list either has not survived or is among the many undated, unlabeled plant lists of Kumlien's that have survived.

Kumlien, who must have been reading one or more newspapers as well as talking about the war with friends and acquaintances, wrote to Fries about the early stage of the Civil War when General George McClellan famously refused to fight: "Our war here is not resolving itself; small skirmishes, many but no decisive battles. Gen. McClellan, the general of the northern army, and the rebellion's General P. G. T. Beauregard have now for many months been entrenched close to one another; both armies together number about 5 to 6 hundred thousand men, but no side dares to attack!" He mentioned the speculation that McClellan would stay put and watch Beauregard, waiting for other Union armies to attack the rear and the flanks of Beauregard's troops. Kumlien believed the main issue to be that "the President has wanted to save the Union and keep slavery according to the constitution, but it is impossible to keep both." Surely, he wrote, the Congress must "abolish slavery, and soon, or else everything will go crazy." Kumlien read that the war was costing the country over two million dollars a day and that the northern states had about six hundred thousand soldiers—"or rather volunteers, spread out across a great number of places." He passed on more speculation about what McClellan might do, concluding, "No matter how the war will end it is expected that slavery has received its death sentence in North America!"[39]

In January 1862, Kumlien received a request for bird skins from the director of the Natural History Museum at Leiden (or Leyden) in Holland. Encouraged by his collecting and correspondence with Fries and the Swedish museum, Kumlien had initiated contact with German ornithologist Hermann Schlegel in December 1861, offering his services in collecting zoological specimens including "fishes mounted or in Alcohol, land and fresh water shells, turtles, crustacean and insects." As to his ability to properly collect such specimens, Kumlien referred Schlegel to the three Swedish professors for whom he was then collecting and "all the ornithologists in the United States."[40] He wrote the same letter to John Edward Gray, keeper of zoology at the British Museum in London.

Early in 1862, Schlegel responded, requesting 128 bird skins and a long list of mammals from Kumlien. To help him fulfill this enormous order, Kumlien hired a man named Andrew Mortonson. At the end of the summer, they packed the specimens in large dry goods boxes and sent them to the consulate general of the Netherlands, who shipped them to Holland to the Royal Museum of Leyden. Kumlien was paid $132 in American money for this order, but not until March 1863.[41] Schlegel placed several even larger orders in the following years, but Kumlien had to wait sometimes three or four years to be paid. His son Theodore said of those years, "When anyone in the family wanted anything, he was told, 'You will have to wait until the money comes from Leyden.'" The four Kumlien children each had lists of things they wanted when the money came from Leyden. When money for one shipment finally came, Kumlien's first purchase was "a long black cloak and a white scarf for Christine."[42]

Though the battles of the Civil War were, of course, not fought on Wisconsin soil, Kumlien knew men who were soldiers, and the horrors of slavery particularly intruded into his mind so that he followed the war as he would follow no other national event.

This daguerreotype of the Kumlien family ca. 1860 shows (top row, left to right) Christine (about forty), Ludwig (about seven), Thure (about forty-one), and Theodore (about five). Svea (about three) is holding a toy dove. Frithiof, not pictured, would have been an infant in arms. COURTESY OF GREGG KUMLIEN

In the early 1860s, Kumlien expressed more than once a desire to go north to collect—north of the Wisconsin River and then west to the Mississippi River, to Minnesota, along some tributary of the Mississippi, and then on to Lake Superior. Though around Koshkonong and in the Baraboo Hills he collected fish, algae, insects, birds, mammals, and plants and sent them to Sweden, England, Holland, Boston, and New York, Kumlien was desperate to see new places—to crawl about and find things new to him and new to science. His botany companion, Edward Greene, then a junior at Albion Academy, would get such a chance to see new places, though he would travel to the south and he would go as a soldier.

In August 1862, the nineteen-year-old Edward packed his *Class-Book of Botany* in his knapsack and joined the Janesville-based Thirteenth Wisconsin Volunteer Unit, which both his father, William, and older brother, Charles, had joined in September 1861.[43] His brother Mansir would join the same unit in 1863.[44] Edward's mother, Abby, supported the war effort by, among other things, making an album of the plants Edward pressed and sent to her from Fort Donelson in Tennessee. The album sold for fifty dollars, and the proceeds were used for the relief of sick and wounded soldiers.[45]

On August 18, 1862, having been denied stored food that was promised them, a small number of Dakota Indians attacked the Lower Sioux Agency on the banks of the Minnesota River. Over the next days, a larger group attacked Fort Ridgley and the town of New Ulm, killing more than two hundred settlers and taking an additional two hundred captives. By September 23, the American Indians were defeated and their captives were freed.

These Indian attacks and the rumors of attacks in Wisconsin created "a strange panic" throughout the state.[46] Though Wisconsin was home to just nine thousand Native people compared to its population of eight hundred thousand white people, this panic swept across the state, causing citizens fueled by rumors and racism to abandon their homes and flee east to where they thought they were safe. Certain that the American Indians had been "tampered with by rebel agents," Wisconsin Governor Edward Salomon appealed to the War Department for guns and ammunition to protect white Wisconsinites from completely imaginary "roving bands of Indians."[47] For a short time, the Wisconsin Twenty-Fifth Regiment was sent to Minnesota as a show of force against any possible Indian "uprising." It was an unsettling time in Wisconsin and Minnesota.

Because of this unrest, Kumlien's plans to travel across the Mississippi River to collect plants were canceled. He wrote Fries that the Minnesota Indians were "enraged and encouraged by the rebels and their agents, so a small army was sent there to pacify the uproar." He accurately reported that thirty-three Indians were hanged in Minnesota, and then he added misinformation: "still it is said they are agitated.—Foolish savages! It will end with their complete extermination!"[48]

If there was panic in Edward Greene's world, he didn't write about it. He was behind the lines with the Wisconsin Thirteenth. On September 10, 1862, he wrote to his "Dear Friend Kumlien" from Dover, Tennessee: "I am seated in my tent this afternoon to write you a few lines to inform you that I am in the land of the living away down in Tennessee. I have been on the march for most of the time since I arrived in Dixie." He reported that the artillery of the Wisconsin Thirteenth attacked the rebels at Clarksville, but that he was not close to the line of fire. He didn't know how many rebels were killed, but "the negroes who were present tell us that . . . one hundred were killed or wounded." Greene told about mangled rebel bodies found in a peach orchard, then abruptly changed the subject: "I have seen a great many new things in the vegetable world since I left home."[49]

Though Greene would remain in the army until the end of the war, it was the "vegetable world" that he wrote about to Kumlien for the next three years. In letters to Kumlien from Tennessee, Kentucky, and Alabama, Greene sent plant descriptions, seeds, pressed plants, and plant lists. He also sent plants to his mother for Kumlien to pick up. He told Kumlien when his mother's cacti were blooming and encouraged Kumlien to go look at them. As Greene grew into manhood and into a botanist, he needed Kumlien's approval (as well as his correct plant identifications), and he wanted to share the beautiful new plants he saw. Though Kumlien seldom wrote back, Greene wrote him long, graceful, plant-filled letters until he returned from the war.

Greene's studies at Albion Academy would have been similar to Kumlien's studies at Skara and at Uppsala, though perhaps not as formal or intensive. Greene, too, studied Latin and Greek, philosophy and literature and history, along with natural history. Later in life, he was described as a refined and private man who liked to botanize alone, a man with an acute sense of smell and a fine sense of humor.[50] His letters demonstrated his sophisticated appreciation of beauty and attention to spiritual values. Kumlien and Greene shared much more than their interest in botany.

From Dover, Tennessee, that September, Greene sent Kumlien new ferns he had found on rocky banks of the Cumberland River: "Some of them are small and beautiful and one very interesting one is called the walking fern." He remarked on the "pretty Labiates" in the mint family, which were still flowering, and the many species of *Desmodium*, or tick trefoil, in the area. These southern forests, he wrote, "must be delightful in early summer," and he listed the Latin names of ten trees entirely new to him, including two varieties of oak.[51] The seeds and pressed plants Greene sent to Kumlien may have been among the ones Kumlien then sent off to Fries in Sweden.

On one "lonesome, rainy day" in the fall of 1862, Greene wrote from Fort Henry, enclosing a photograph of himself. In Tennessee, the days were still warm and Greene saw a *Chelone*, or turtlehead, shorter and darker purple than the Wisconsin variety, still blooming. He also described his first cypress trees, whose "stumps arise from the roots, some to the height of three feet, but of all sizes . . . like a group of small mountains."[52]

In November, Greene sent Kumlien a package of seeds, acorns, and pressed plants, and he described a beautiful road along the Cumberland River, which passed through a deep ravine where a clear stream arose from a large spring. "The rocks on its edge are covered with mosses and higher up the ferns are still green and beautiful," Greene wrote. "I thought as we marched along that it would be a beautiful place in May or June and doubt not there would be many new things there for you and me."[53] He enclosed a little grass that grew along the river. Kumlien seldom responded to these poetic letters from his young friend Greene, but he carefully saved them.

Kumlien apparently saved his energy to write letters to his professor, but these were filled with news of the war. On February 10, 1863, Kumlien wrote to Fries: "Perhaps the R. Mr. Professor is expecting a few words about the ongoing war? There was a time when I thought that the North and justice would win over slavery, but now I am starting to fear that this nation still hasn't been punished enough for its sins towards the poor blacks!" Then he wrote about a "secret organization which exists here, which supports the South and consists of the most deplorable, brutal, and ignorant people. It has grown large and is numerous, and they are now doing everything to hinder the government's measures and to help the South." Kumlien may have been referring to the conservative Democrats, called Copperheads, often German and Irish immigrants who were fiercely opposed to the Emancipation Proclamation. He went on to complain about the northern generals and what he saw as their inaction and corruption. "I am worn out by it all," he admitted, telling Fries that he didn't think the Republican form of government served America. All of America, he wrote, "is about getting 'the almighty dollar.'" He thought real patriotism was rare. If he had the means, he would go back to "old honest Sweden." He hoped that the Emancipation Proclamation would have an effect, but struggled to stay optimistic. He said he was "hoping for the best, but right now it looks dark."[54]

The America of the early 1860s was not the one Kumlien had envisioned in Sweden. He was disillusioned by the war, by the Indian attacks in Minnesota, by Americans' pursuit of the "almighty dollar," by the disappearance to agriculture of the natural world he had seen in 1843. What Kumlien saw valued by those around him was not what he valued—the beauties of the natural world, the wonders of scientific knowledge, music and art and literature. And a spiritual life. Those were the same values held by Edward Greene. But in the years of the Civil War, Kumlien seemed to have his eyes toward the east and Fries more than to the south where his young friend was a soldier.

Among the plants Kumlien sent Fries in the spring of 1863 package were ferns he had received from "a young botanist, who is a soldier and who for a while has been stationed in the state of Tennessee, at Fort Henry." Kumlien said that as soon as the war ended and a northerner could safely travel in the South, he would go there to collect because "there are lovely things!" He sent Fries "*Aster No 34* which should be named *pinifolius* or perhaps rather *abietifolius*, if such a name can be sufficient. Can it be the same as No 11 in the small bundle and No 12 . . . from Tennessee??"

In February 1863, Kumlien wrote to Fries that if everything turned out well, he planned to spend more time in the coming season on plant collecting, with a particular focus on *Carices*, or sedges. He added, "If the Indians mind their manners, I will cross the Mississippi River a short way, for there I think I can find something that doesn't exist here." Kumlien also promised to send dried lichens to Europe: "A nuisance, due to the war, is that paper is now more than twice as expensive as it was before, but it can't be helped. I have decided to collect plants this year and unless something unexpected comes up, I will do it."[55]

Greene wrote five letters to Kumlien in the spring of 1863. Kumlien wrote back twice, both times asking him to collect zoological specimens that Kumlien could send on to Fries in Sweden. From the tone of Greene's responses to these requests, it's clear that he didn't really want to collect animals. He didn't have the knowledge, though his father got him a book, or the equipment. Carrying animal specimens when he was on the march would have been difficult, and Greene was not used to killing birds or mammals the way Kumlien was. He did it—at least, a little—because his friend asked. But it was a lot for Kumlien to ask of Greene. Perhaps Kumlien thought the scientific value of the specimens would outweigh any personal objections Greene might have to collecting them.

On March 12, 1863, Greene wrote from Fort Donelson excusing Kumlien for not answering every letter. Greene knew that Kumlien was "otherwise engaged." He was glad Kumlien was ready to begin the labors of another season and wished him success, though Greene regretted that he could not "share the pleasure of roaming about the prairies, woods, and marshes of old Wisconsin with you." He described the romantically situated white-flowering hepatica growing in crevices of rock along the Cumberland River. And he told Kumlien that he had written to Alphonso Wood, author of the *Class-Book of Botany*, which they both used. Wood responded to Greene saying that he, too, had "tramped about" that rich field for botanical research along the Cumberland River "with box, press, spade and compass."[56] In May, Greene wrote Kumlien that Wood had sent him a new and greatly improved edition of his *Class-Book of Botany* with better descriptions and with the number of genera greatly reduced. The new edition was said to include all known plants in the United States except those of south Florida and west of the Mississippi, but Greene told Kumlien that he found an *Allium* (in the onion family) and *Erigeron*, daisy fleabane, that he thought were not in the book.[57]

By 1863, Kumlien's post office address had changed from Albion to Busseyville in Sumner Township, Jefferson County, Wisconsin. Sumner Township had been formed in 1858 when Koshkonong Township was divided. Busseyville was the site of the sawmill Sven Bjorkander had owned on Koshkonong Creek. English settler Thomas Bussey had established a grist and flour mill at the same site. In the early days of his life in Wisconsin, Kumlien had to walk ten miles to Fort Atkinson to pick up mail and, more recently, he walked four miles to Albion. Now there was a post office about a mile and a half from his homestead.

One day in June 1863, when Kumlien was returning from a short excursion to look for *Arethusa*, which he later described to Fries as "modestly stick[ing] its beautiful head up while half-tucked away in loose turves of moss," he stopped in Busseyville and picked up a letter from Fries. He read that Fries's wife, Christina, had died and that Fries was not in good health. Kumlien remembered "Mrs. Prof. Fries," though it had been more than twenty years since he had seen her, and he sent his heartfelt condolences.[58]

Elias Fries was an old man when he sent this photograph of himself to Kumlien in the 1860s. UPPSALA UNIVERSITY LIBRARY

In his letter, Fries had apparently told Kumlien he was stepping down
from his position at the university because of ill health. He told Kumlien
that his son would classify and distribute any lichens that Kumlien sent.
Was this son, Kumlien asked in his response, Fries's oldest son, Thore,
whom Kumlien remembered as thirteen or fourteen years old? Or was it
one of the younger of Fries's six children? Kumlien also asked who was to
be the "Mr. Professor's" successor?[59] Was it any of his old Uppsala comrades
or acquaintances, who were now "men of the State, the former boys"?[60]

In his letter, Kumlien added that he was about to arrange a large collection of sedges and that he planned to walk north one hundred English miles that season to collect stone lichens, miterworts, partridgeberries, and dragonflies.

In July 1863, Greene wrote Kumlien again from Fort Donelson describing more than twenty species of plants he had collected and pressed. Greene's experience of the war was, to say the least, not typical. He said he enjoyed his soldier life very well and had seen many beautiful things in nature, though he was now ready to return to Wisconsin and for the war to be brought to an end. "If our armies gain many more such glorious victories," he wrote, "I think we may hope for and expect peace soon. . . . Perhaps I shall be disappointed but I think not."[61]

Greene would be disappointed. The bloody war would go on for almost two more years. Greene closed his July letter to Kumlien, "P.S. Please write as soon as convenient."[62] Apparently, it wasn't convenient until the end of the year, though Kumlien did write a letter to Professor Fries first.

Fries had sent Kumlien a note of twenty pounds sterling, and Kumlien, on August 25, 1863, was sending a receipt. He asked Fries to please identify "some seeds from Tennessee, and some *Aster* and *Carices*." *Aster amethyst* he thought must be rare, as he found only two stands of it. The other little aster, which Kumlien reminded Fries he had "so kindly" named after Kumlien in 1860, "is only to be found on a sand hill around here and it is rather rare."[63] Angie Kumlien Main remembered the Kumlien aster as "blooming over the fence near his home on the Benneworth place in a ditch by the old turnpike which still crosses the marsh."[64]

Kumlien, in this August 1863 letter, told Fries he wanted to travel to Oregon for botanical and zoological specimens, but he knew he could never afford it. He had sent bird skins in a recent shipment to the Swedish Museum of Natural History, but he hadn't heard if his large shipment arrived. Like Greene, Kumlien believed that the war was ending. "Charleston will be falling soon if it has not already," he wrote, "and then there isn't much of value left to take."[65]

After the extreme cold of the long winter, 1864 for Kumlien was a quiet year of waiting for the war to be over. Greene didn't write to Kumlien for all of that year. Kumlien apparently wrote to Greene in November, but we don't have this letter. There were no letters to Brewer that year. And none to Fries.

But that year Kumlien was busy with collecting. On March 26, 1864, Kumlien made a plant list with seventy-six items, many of them difficult-to-identify, grasslike sedges. Kumlien sent a shipment of specimens to the Zoological Museum of Berlin in February and, in April, a large box of specimens to the Royal Museum at Leyden in the Netherlands. Also in 1864, Kumlien began collecting for oologist Edward A. Samuels in the agriculture department of the state of Massachusetts.

On January 4, 1865, Greene wrote to Kumlien for the first time in almost a year from Huntsville, Alabama. Since he last wrote, he said, he had not seen active fighting, "what the ancients call the 'gloria belli,'" or the glory of battle. But Company K had suffered from sore feet, weary legs, and short provisions. And Greene had lost most of his specimens when his company was ordered to leave Claysville, Alabama, where they had been patrolling along the Tennessee River. Leaving in a hurry, they had to burn "a good many valuable things"[66] to keep them from falling into enemy hands. But Greene was able to jettison some clothing and make room in his knapsack for most of his plant specimens.

Because we have only Greene's half of this correspondence, we have only echoes of what Kumlien wrote about his religious beliefs in a letter on November 13, 1864—something that, as far as we can tell, he wrote to no one else. Greene wrote about wanting to live in the south or west of the Mississippi, but then stated that he would "cease from building these airy castles for the future since it is so impossible as you say, for us to know what it will realize to us. I suppose we who acknowledge our dependence upon the merits of our Lord and Savior for our hope of eternal happiness should be more willing to submit—to do his will when, where, and however he may see fit, than to accomplish our own plans."[67] Responding to Kumlien's letter, Greene wrote, "We cannot exercise proper faith in him unless we are willing and ready to prefer what may seem to be his will, to our own. . . . I am glad to perceive by your writing that your hopes of eternal felicity, beyond this scene of strife and mortality, are centered in the merits of the crucified Son of God, which I too believe to be the only hope, for fallen, and sinful humanity."[68] Though Kumlien had joined the Episcopalian church of Gustaf Unonius and had his children baptized, he almost never put his Christian beliefs in writing. It is not surprising that the spiritual and religious Greene, a poetic young man whom Kumlien had known since Greene's childhood, would be trusted, particularly in wartime, with these rare expressions of religious belief. Greene concluded his letter by telling Kumlien about some uncommon lichens he found on cedar trees on the bank of the Tennessee River: "a species entirely new to me."[69]

On March 5, 1865, Kumlien apparently wrote to Greene saying he was going to Kansas. His little ones had been unwell, but they were recovering. Greene, on March 18, thanked Kumlien for his kind admonition with regard to abiding in close intimacy with God our Savior: "I am glad for your sake and for her sake and for the sake of God that you have a sister so able to give you council and cheer in the comforts of our holy religion. . . . I say *our* religion for it is all the same in every fundamental principle and I with you can see no reason why you need leave the communion of the church into which you were baptized."[70] We can only speculate that Christine and Thure were together considering leaving Unonius's congregation for the Lutheran Church of their youth.

On April 9, 1865, General Robert E. Lee surrendered to General Ulysses S. Grant at Appomattox Court House in Virginia, and the war was over. But it took weeks for the news of the end of the war, and then of the April 15 assassination of President Lincoln, to reach the soldiers in the field. Kumlien apparently wrote to Greene on April 23, as Greene responded on May 2 by writing, "You express the same sentiment with regard to the death of Mr. Lincoln which is felt by us here. We received the terrible tidings while at Jonesboro, but were slow to believe such a horrible report."[71]

"Evidently," Greene wrote, "the war is ended and every move seems now to indicate a speedy dismissal of a large number of troops."[72] If Kumlien did not go to Kansas that year, Greene suggested that perhaps he could go with Kumlien the following year.

One night while he was in Huntsville, Greene had dreamed of mountains: "lofty mountains whose sides were clothed with the richest verdure of various kinds of trees and whose summits rose above the floating clouds. . . . I endeavored to ascend one, but it was so steep that I became extremely tired and then I awoke from my vision" in the same place where he had been sleeping for months.[73] But Greene thought that he might partially realize his dream because the company was headed for

mountainous country. At Jonesboro, Tennessee, they camped "in
full and delightful view of the great
Smoky Mountains of North Caro-
line." Greene wrote Kumlien, "You
will easily imagine with what long-
ings of heart this child used to stand
and gaze at those sublime but rugged heights."[74] In the same letter,
Greene wrote that he had climbed
to the top of Lookout Mountain in
Tennessee where he found a few
flowers he had never seen anywhere
else and where he collected a dry and
fragile lichen for his friend Kumlien. From 1870 to 1875, Greene's
work as a botanist would take place
in the mountains of Colorado.

Edward Lee Greene likely sent this
photograph of himself to Kumlien from
Colorado soon after Greene had been
ordained as a minister in the Episcopal
Church in 1873. WHI IMAGE ID 147768

Usually, Greene closed his let-
ters to Kumlien, "Your friend, Edward." But Greene signed his letter
of May 2, 1865, his last as a soldier: "With best wishes to you and your
family, I subscribe myself, with love Your friend, Edward Lee Greene."
Hearts were sore and emotions were more openly expressed at the end of
the terrible war and at the terrible loss of President Lincoln.

The end of the war marked the end of one the most intense collecting
periods of Kumlien's life. And after the war, there were no more letters to
or from Elias Fries. Though Fries did not die until 1878, the ill health that
caused him to step down from his post may have prevented him from writing to his old student.

When Greene returned to Albion Academy to complete his bachelor
of philosophy degree in 1866, he and his friend Kumlien must have had a
few chances to explore the woods and prairie and tamarack swamp near
Lake Koshkonong. With his degree completed, Greene left Wisconsin to
teach school in Monticello, Illinois, from where he wrote to Kumlien once
again. Setting off for the Colorado Territory in 1870, Greene studied mountain flora and studied for the Episcopal priesthood from 1871 to 1873. Then

he became a missionary in 1874, botanizing in the Southwest. He gave up the ministry to take a position as professor of botany at the University of California at Berkeley in 1882. Greene eventually became a Catholic and a professor of botany at the University of Notre Dame in South Bend, Indiana. He died in 1915.

In 1888, when he was one of the most important botanists of his age, especially regarding the plants of the West, Greene said that his interest in the vast and then-unknown flora of the western territories had been awakened by reading the botanical parts of the Pacific Railway Survey Reports with his friend Thure Kumlien. We can picture the dark-haired boy and the light-haired "courtly gentleman" leaning over the pages of this important work, both of them dreaming of searching for plants in mountains higher than the Baraboo Hills.[75]

Drawing of cattails attributed to Thure Kumlien.
COURTESY OF NANCY KUMLIEN-GARRY

12

THE SONS AND SCHOOLS
OF FINE-GRAINED MEN

1866–1875

One day in March 1866, with the ground still frozen and the collecting season not yet begun, Thure Kumlien wrote to Edward A. Samuels in the Massachusetts Department of Agriculture in Boston to tell him that, yes, the little wading bird called the solitary tattler [solitary sandpiper] did breed at Koshkonong, but its nests were difficult to find. Kumlien had collected eggs and nests for two seasons for Samuels, who wrote that he was preparing a "paper on our northern oology." That paper became *The Ornithology and Oology of New England*, a book first published in Boston in 1867.[1] By 1866, Samuels and Kumlien were chummy enough that Kumlien sent him a poem about longing for his childhood home in Sweden, and Samuels asked if he could come and board at the Kumliens' for two or three weeks that summer. Kumlien, living in the log cabin with his wife and four children, told Samuels that he was sorry, but he would probably be in Minnesota or Iowa collecting at that time. However, if Samuels came around and Kumlien wasn't home, eggs could be collected by his boy, Ludwig, "a lad only 13 years old but perfectly able to identify what we can get here."[2] That summer, Samuels never came to Koshkonong, and Kumlien again canceled plans for collecting in the West. But Kumlien's lively, talented, and ambitious sons, as well as his young friend

Rasmus Bjorn Anderson, began to add ease and help and humor, as well as worry and angst, to Kumlien's rich life.

In June 1866, all of the Kumlien family—Thure and Christine, thirteen-year-old Ludwig, eleven-year-old Theodore, nine-year-old Swea, and seven-year-old Frithiof—likely attended the festive graduation, a local holiday, at nearby Albion Academy.[3] A hired band on a wagon played as they came into Albion. Students in new clothes and faculty in their best marched in a procession from the school to a leafy grove, where there were songs, speeches by dignitaries, and original orations by students. Edward Lee Greene was at Albion Academy that year, as was Rasmus Anderson, who grew up on the prairie nearby.

Though an able student at Luther College in Iowa, Anderson had been expelled by the president of the college for being a "weed," spreading discontent and rebellion among students and eventually encouraging them to circulate a petition and threaten a strike.[4] Living conditions were unhealthful, and students were required to chop wood and polish shoes for their teachers. Anderson had tried to reform Luther College, which was later described by Anderson's biographer Lloyd Hustvedt as an institution "where the surroundings were drab, the regimentation severe, and the traditions yet to be made,"[5] but Luther College did not want to be reformed. Now, in the summer of 1866, Anderson sat in the Albion grove and wondered what to do next.

The village of Albion, less than four miles from the Kumliens' cabin, had been founded by well-educated families from New York and Rhode Island. Flexible in their belief system, they were guided by principles of religious freedom, the separation of church and state, and the abolition of slavery. Calling themselves Seventh Day Baptists, because they worshipped on Saturdays, they were people who valued good and affordable education for both women and men.

The people of Albion, led in 1851 by Charles R. Head, drew up the school's building plans, cut its bur oak beams, made its nails in the Albion blacksmith shop, and built its first building with bricks they made of yellow clay found in the village. Bringing young elms and maples from the area near Koshkonong Creek, they planted rows of trees on the campus, watering them with buckets of water carried from the nearby creek.[6] The second and third buildings were built with borrowed money that was soon paid off. The beautiful buildings and campus reflected the value these people placed on education. South Hall, a three-story brick building with twenty-eight rooms used as classrooms and dormitories, was lit by forty-eight twelve-pane windows.

Albion Academy promoted "plain living as well as high thinking."[7] Though called an academy, it was a four-year college offering a bachelor's degree and courses in Latin, Greek, German, French, mathematics, history, English literature, and the sciences. Albion Academy was the first educational institution in Wisconsin open to both men and women on an equal basis.[8] And with two hundred students in the 1860s, it was the largest institution of higher education in Wisconsin—larger, even, than the university in Madison.[9] With rates for tuition and room and board kept extremely low, Albion Academy became "an educational Mecca" for poor students, who boarded on campus or lived with local people.[10]

Professor A. R. Cornwall, a graduate of Union College in New York, served as Albion's principal from 1861 to 1878, a period of increasing attendance and vigorous study. Cornwall, a polished classical scholar of Latin and literature, was also enthusiastic about the study of natural history and one of Kumlien's close friends. Cornwall brought Albion students out to Kumlien's home to see his collections of birds and plants and to hear him talk on nature.[11] Until they had a falling out over Anderson, the Cornwalls and the Kumliens were said by Angie Kumlien Main to have visited each other's homes every few weeks.

By the mid-1860s, Kumlien was known at several nearby institutions of higher learning as an expert on natural history. Perhaps as early as 1865, Kumlien had prepared a cabinet of mounted birds for Albion Academy, and in August 1866, he was invited to speak at Milton Academy. Taking thirteen-year-old Ludwig with him, Kumlien stayed at a hotel in Milton, bringing along mounted birds for the academy's museum.[12] But it was in 1867—through the influence of Anderson—that Kumlien was offered a position on the faculty of Albion Academy.

Young Anderson was gifted in intellect, attractive in person, and outsized in ego. Growing up in a Quaker-influenced Norwegian immigrant home near Albion, he frequently visited the Kumlien home as a boy, sitting at the feet of the Swedes as they recited and sang the works of the Swedish poets Esaias Tegnér, Johan Ludvig Runeberg, and Erik Gustaf Geijer. This

President of Albion Academy A. R. Cornwall (left) and his wife (right) were good friends of Thure and Christine Kumlien, though the couples had a falling out in the late 1860s. Later, after the Cornwalls had moved away, they wrote with affection to Thure. WHI IMAGE ID 147772 AND 147773

experience fostered in him a love and appreciation for Swedish literature and history.

The day he sat in the grove at the Albion graduation in 1866, Anderson wrote that it occurred to him that Albion Academy might be able to further the enlightened causes of the Norwegian-American Educational Society, which he had helped found in Iowa. Soon after the 1866 graduation, the personable Anderson convinced the Albion Academy board that if they hired him to teach, he would attract many Norwegian students to Albion.

Once he joined the faculty, Anderson became an increasingly vocal and public figure, writing articles for the new Norwegian newspaper *Skandinaven* and traveling around Wisconsin to recruit—successfully—Norwegian students for Albion Academy. Exaggerating his personal success as he recruited and raised money, Anderson and his immense ego began to chafe under the authority of Cornwall and the Albion board. Anderson had good ideas for promoting the educational and cultural life of Scandinavian immigrants and the common schools. He was a gifted man who achieved much in his long life, including founding the first department of Scandinavian studies in the country at the University of Wisconsin in Madison and becoming the US ambassador to Denmark, but he was as flawed as he was gifted.

Yet the brilliant Rasmus Anderson must have been an exciting friend and a thrilling conversationalist. With his command of several languages, his love of literature and music, and his passion for the rights of students, this Romantic figure may have reminded Kumlien of his student days at Uppsala. Their friendship elicited from Kumlien some of the most revealing letters of his life. The two men were generally frank with each other, yet each, to some degree, romanticized the other. Anderson's description of Kumlien as a "charming, scholarly, fine-grained man" seems both romantic and accurate, revealing Anderson's facility with language. "Fine-grained" is a poetic, right-seeming descriptor of Kumlien's sensibility and temperament, the acuity of his observations, and the refinement of his expression and emotion.

In his memoir, Anderson claims a good deal of credit for Kumlien being hired at Albion Academy. Anderson says that after President Paul Ansel Chadbourne of the University of Wisconsin at Madison heard what a good job Anderson was doing teaching and recruiting at Albion, Chadbourne offered Anderson a "larger field in which to operate" at the university.[13] Anderson accepted the offer and tendered his resignation at Albion, but the Albion board refused to accept it: "At Albion I could easily, they asserted[,] be the first, while at Madison and at the university my identity would be lost."[14] Rasmus relented. But, not one to waste a good bargaining position, he made several demands of Albion. One was that four prominent Norwegians be placed on Albion Academy's board; the other was that "T. L. Kumlien should be dragged out from his obscurity . . . and given a regular professorship of natural history at Albion academy."[15] Both were agreed to, he stated in his *Life Story*.

But Anderson's accounts were never the whole story. Lloyd Hustvedt, in his biography of Rasmus Anderson, told it differently. Having no intention of staying at Albion Academy, Anderson was looking for employment opportunities at several places, including Yale and the university in Madison. Hustvedt believed that it was Anderson's friend—and

When this photograph was taken in 1875, Rasmus Bjorn Anderson was at the beginning of his career at the University of Wisconsin. WHI IMAGE ID 147770

Kumlien's—Edward Lee Greene who put Anderson in touch with President Chadbourne. Chadbourne offered Anderson a position as either a postgraduate tutor or instructor of modern languages, despite the fact that Anderson would be at the university working on his bachelor's degree. All accounts agree that when students at Albion learned that Anderson was planning to leave for Madison, twenty-six Norwegian students signed a petition asking that he stay. President Cornwall, at Albion, also wanted him to stay. Anderson, from his good bargaining position, did get the four Norwegians appointed to the board, and, somewhat unbelievably, he did arrange matters so that he and President Cornwall would share the financial affairs of Albion Academy. Anderson said that Kumlien was brought in at his insistence at this time, though Kumlien likely had a part-time or temporary position at Albion before this. Because Albion Academy records were lost in a fire, we cannot know exactly when Kumlien joined the faculty. Hustvedt asserted that Kumlien was on the faculty "through Anderson's efforts" by 1867, rather than hauled from obscurity, as Anderson claimed, in 1868.[16]

However and whenever they got there, both Kumlien and Anderson were instructors on the faculty at Albion Academy for the momentous 1868–1869 school year. Kumlien was a popular teacher of Latin, zoology, and botany. He added to and maintained the academy's cabinet of natural history specimens. And he was the naturalist to whom students and faculty brought oddities. One day, one of Kumlien's students shot a strange bird on the Albion campus and brought it to Kumlien to identify. It was the first documented appearance in south central Wisconsin of the invasive English sparrow.[17] Kumlien had not seen the English sparrow since living in Sweden.

In the summer of 1868, having had some success raising money and recruiting Norwegian students for the school, as well as having been elected an alternate delegate to the Republican national convention and securing a friendship with famous Norwegian violinist Ole Bull, Anderson went into the school year with an exaggerated sense of his importance and power. Right away, he began to plot against Cornwall for control of the school. Anderson wrote to his brother in January 1869 that he intended to get a professorship at Albion. His plan was to "burst this school at the close of the present term. . . . I am a great schemer and

Thure Kumlien in 1868. WHI IMAGE ID 147680

you know it."[18] In February, the Albion board of trustees learned of his plot and publicly dismissed him. Though it's hard to tell from this distance, especially with the loss of records, Anderson may have had a point. Cornwall had lost the confidence of some of the faculty and students, perhaps because his financial records were found wanting. When Anderson was dismissed from Albion Academy in February, Kumlien resigned from Albion in protest: "I desire to have it definitely understood that I am no longer connected with Albion Academy, that I very much disapprove of the course pursued against Professor Anderson by the managers of the school and can give the institution no kind of support until it has satisfactorily adjusted the difficulties pended."[19] In late March 1869, Kumlien returned to Albion to collect his things. His teaching career at Albion was over—for a time.

When Anderson was dismissed from Albion, he was a married man and soon to be a father. He again wrote President Chadbourne at the university for a teaching position. Chadbourne's response was cool, but he conceded that Anderson could come see him to explain what happened at Albion. Anderson took the next train to Madison and, persuasive man that he was, convinced Chadbourne to take him on at the university. Anderson immediately moved into a friend's house in Madison, leaving his pregnant wife to stay with her parents in Cambridge, Wisconsin.

On March 21, 1869, after he'd had time to think over the events of the previous months, Kumlien wrote a long letter to his "Friend Rasmus" in Madison. Anderson needed, Kumlien wrote, an insightful and honest friend who would tell the impetuous young genius to leave off revenge and scheming and newspaper article writing. Kumlien encouraged Anderson to use his brains for the taking *in* of information: "In a word mind your own business for a while and let others alone. . . . Do not tell everybody all you know and all you think because it will be your ruin."[20]

Kumlien acknowledged that Anderson might be offended by his letter, but, he said, "I am probably more a real friend than any you have." Kumlien concluded this letter with a reminder that he had known Anderson since he was a boy. "If you get offended at me now," Kumlien wrote, "I will pinch your nose when I see you next."[21]

A few weeks later, Kumlien wrote Anderson again about the "disagreeablenesses in Albion."[22] He said, "I am going to talk a long while and I beg you not to be tired and above all not to be angry with me, because what I am about to say is prompted by the very best feeling and, will you believe it, by a stronger friendship than you yet have from anywhere."[23] The previous fall, Kumlien had found that Anderson possessed a "monstrous great ambition to rule . . . and required that every one . . . should do [Anderson's] bidding and woe to him that dared hold an opinion of his own." Kumlien then repeated some of the points from his previous letter and added that Anderson shouldn't have brought Edward Lee Greene into the controversy. Kumlien thought that Anderson should, by "exemplary conduct, uncommon diligence & by *dashing excellence as a student*, raise far above" his critics.[24]

Kumlien continued, "Now Rasmus I know that . . . I speak to a man that has a very great opinion of himself and one who thinks it next door to impossible that he can be in any way wrong." But, he added, "at the same time there is stuff in you for a *great man*, a *useful man* and possibly a *good man* if only that wonderful mind of yours could be trimmed a little yes considerably."[25]

Kumlien mentioned the Albion "fuss" in his own life as "a very unpleasant dream." He hated to go to Albion; the thought of the place made him feel "sickish." He could, he said, live without those people at Albion. Kumlien must have broken completely—though temporarily—with his friend

Cornwall. Kumlien's wish was for "peace and quiet and I hope that I will find it. I hope so—yes I do." "Write to me soon," Kumlien concluded, "but don't scold me too much for my insolent letter."[26]

Anderson certainly did scold. He disowned his friend in a letter Kumlien received one day after sending his, and "a sorrowful night it made for me," he wrote to Anderson five days later.[27]

"So bidding me goodbye are you?" Kumlien wrote. "No not quite so fast Rasmus." We don't have Anderson's letter, but we can gather that he probably accused Kumlien of neutrality as well as siding with Cornwall and other enemies of Anderson. Kumlien explained that he could have written Anderson a letter of praise for all he had done for the institution of Albion and "for me poor devil." But, he asked, what would be the good of that? "From the first time we were together in Albion," Kumlien wrote, "you very carelessly allowed me to assume the role of a Damper and I, old fool, have now held the idea that an occasional dampening was good for you, of course deeming the period of last fall & winter I could not very well do so to my superior." These observations of his role as a "damper" to Anderson's excesses, as well as Kumlien's sense of what was not appropriate when Anderson was his superior, demonstrate Kumlien's more "fine-grained" knowledge of human behavior.

Backing off from his stern tone, Kumlien said his previous letter was "lamentable" and that it was written by "an old silly man" who meant well instead of evil, a man in whom there was no "duplicity, falseness nor treachery." He described Anderson as "the greatest Scandinavian in educational matters in the United States" and as "a great man in embryo."

Then Kumlien, somewhat surprisingly, admitted he may have misunderstood Anderson. He wrote: "I am ready on my bare knees to ask your forgiveness." Kumlien must have felt he had to do this to heal the broken friendship. He said he meant as well as "a father can with his son or a brother with a brother." Kumlien felt so awful, he said he would never "seek any place as a public man" but would instead close himself into his "old log shell." He said he couldn't bear fuss of any kind. For the rest of the long letter, Kumlien flattered, explained, cajoled, joked, and expressed his affection for Anderson.

There is a genuine bended-knee apology in this letter. But there's more to it than that. Kumlien refused to accept Anderson's rejection of him. The friendship meant too much to Kumlien. He set out to talk his way back into Anderson's good graces. And he succeeded.

Soon all was normal between the two. In July, Kumlien was congratulating Anderson on his new position as professor at the university, hoping Anderson and his wife and new baby daughter could come for a visit, and passing on gossip Kumlien had heard about Albion Academy. Anderson had lent Kumlien a book by Chadbourne, probably his 1860 *Lectures on Natural History*, which Kumlien wanted to keep a while and read again. In the fall, Ludwig went to Madison and stayed with Anderson and his family for a time. Ludwig wanted to attend the university, but Kumlien wrote Anderson in October that he could not "see any possibility in it under the present circumstances." Kumlien was in debt, a debt that he said he "must pay and shall pay if the boys have to help me to do so."[28] Ludwig was soon working as a clerk in nearby Cambridge.

On the last day of December 1869, Kumlien wrote to Anderson on a subject that, as far as we can tell, he wrote to no one else. Kumlien said he "had the blues very bad, perhaps harder than I ever had them; I have not got over them quite yet but the attacks don't come so often and . . . I have my natural color."[29] Though we can't be sure about the cause of Kumlien's blues, he ascribed them to his money problems. Yet it's worth noting that almost one year earlier, in January 1869, John Cassin had died of arsenic poisoning from the powders used on the many bird specimens he handled.

Whether or not arsenic was a factor, Kumlien was in debt—enough to cause the blues. And he had been stiffed by a collector named Anderbrook for whom he'd collected eggs and birds skins the previous summer. All that work, he wrote Thomas Brewer, was a loss: "Nice discovery for a man in debt and no means to pay."[30] Brewer, ever the helpful friend, got his agent and his lawyer on the case at no charge to Kumlien. Kumlien then sent Brewer a package of the correspondence between him and Anderbrook. Brewer wrote that he sent the package on to Phelps, the New York agent, who passed it on to Simpson, the lawyer. Brewer, thoroughly enjoying the detective work, wrote Kumlien that Anderbrook was a "hard customer" who didn't seem to exist at the hotel where he said he lived, so Brewer sent Phelps to try to apprehend him. "We will keep a sharp look out for the fellow," Brewer promised, "and sooner or later he must turn up and rather than go to Sing-Sing he may pay you your $100 though it looks now rather dark."[31] We don't know if the money was ever recovered. It still looks rather dark.

Brewer's health had begun to fail as he suffered from "sciatic rheumatism," which he said affected his memory "so that I cannot always correctly recall what I have done or which omitted to do."[32] Brewer was fifty-seven, three years younger than Kumlien. His handwriting was becoming more difficult to read. But he worried about Kumlien. After the Great Chicago Fire and the Peshtigo Fire in northeastern Wisconsin, both of which started on October 8, 1871, Brewer was relieved when he finally heard from Kumlien in December. He had begun to believe Kumlien had been "burnt out."[33] But the following year in November, it was Brewer who was burnt out in the Great Boston Fire of 1872. Though he had to move his publishing business from Milk Street, he reported to Kumlien that he had "lost nothing ornithologically."[34]

On January 1, 1872, Kumlien wrote to his "Friend Rasmus," now a professor at the university. Instead of leaving his calling card at a professor's door on New Year's morning the way the Uppsala University students used to do, Kumlien was writing a "few lines" to Professor Anderson expressing his New Year's wishes. The few lines became a six-page letter, all written in the playful and intimate voice of Kumlien's found nowhere but in letters to Anderson. Kumlien had hoped to see Anderson when he came to the Koshkonong Prairie on his vacation. But perhaps Anderson didn't get to the prairie on his vacation, perhaps he wasn't even at Lina's wedding. We don't know who Lina was and Kumlien didn't attend her much-talked-about wedding either, but he hoped Anderson was there: "If you was not you certainly ought to have been you know. You see I have been told that she is married and that there was a great doing, some 2 or 300 people being invited."[35] Kumlien said that he hadn't been farther from home than Albion since he was last in Madison.

Then, in an awkward shift of subject, he wrote: "Now speaking of Madison—my thoughts naturally swing to my last attempt to get engaged there." This seemed to be his real reason for writing. However, Kumlien was at this time apparently employed again at Albion Academy teaching botany, ornithology, and zoology. At Albion, he was recognized and appreciated as a good teacher, yet he sought what he must have seen as a higher status or higher-paying position at the university in Madison. Anderson had apparently told Kumlien that "Rev. Doctor," probably John Hanson Twombly, the new president of the university as of 1871, had suggested Kumlien write to him about a position there. Kumlien did write, but then, when there was no response, regretted his letter. He blamed the nonresponse on "this epistle of mine," which he called "the very essence of silliness, the most sublime extract of stupidity or the roughest outpourings of a simpleton or an illiterate ass."[36]

An undated draft of a Kumlien letter to Twombly exists, but there seems to be nothing in it that Kumlien would be embarrassed about. It's a pretty good job application letter, beginning with his first line: "My object in seeing you the other day was to be engaged at the University to take care of and increase the collections of Nat. History." He said he didn't like to speak favorably about himself, which he would rather see done by others, "if I deserve it." He gave a bit of history of his education and collecting in Sweden and listed the European naturalists he had collected for. Interestingly, Kumlien had unsuccessfully applied to work at the University of Wisconsin during the reign of three of the first five presidents. He had previously contacted John Hiram Lathrop, president from 1849 to 1858; Paul Chadbourne, president from 1867 to 1870; and now Twombly, the fifth president of the university. Chadbourne had told him that the means were wanting, so "that was the end of it." Which was too bad. Kumlien's best chance at the university was probably with Chadbourne, the naturalist.

In the application letter to Twombly, Kumlien wrote that the university's "Cabinet, certainly not very large, stands precisely as it did years ago, nothing added and I presume nothing abstracted." Kumlien believed that at a university, it was "disgraceful . . . that a number of the birds are wrongly named." He pointed out other blunders in the labeling: "For instance a species of the North European [now Eurasian] Jay (*Garrulus glandarius*) is given as a Lanius from South America. Of two woodpeckers of the same species one is named right and the other wrong. Among the specimens in the Cabinet are some rarer birds that should be taken care of and some very poor ones of common birds ought to be thrown away and replaced with good ones."[37] Clearly, Kumlien had spent time at this cabinet over the years, watching as the mislabeling was not fixed and the sad old common birds got sadder. Perhaps it was this last paragraph, critiquing the existing cabinet, that he regretted.

But Twombly was busy advocating for the education of women at the university, which the regents were against, and he resigned two years later. Kumlien wrote to Anderson on the first day of 1872, "It seems to me that, after he had told me to write, some kind of an answer or communication in due time would have been . . . in order and that the non answering must mean nothing more nor less than contempt or insult." Not very convincingly, Kumlien said, "I care nothing for that now—but it did prick me a little at the time—I own that—but not now."[38] In the middle of his fifth page about this disappointment, Kumlien wrote, "but it's of no use to trouble my mind about that more—I have sense enough to . . . take it cool and try to make up my mind to think of the matter as little and as seldom as possible." This would be, he said to Anderson, his last attempt to get a position at the university.

Kumlien closed this New Year's Day letter by saying that his health had been good "with exception of some gentle attacks of pleurisy, caused by carelessness, and some cough." Was he careless with the arsenic powder? Christine was not well. Ludwig—who would have liked to be at the university, too—was busy during the day chopping wood and hauling it to Albion and in the evenings was "busy with his pencil, but soon will have to strip our tobacco."[39]

Anderson visited Kumlien in March 1872, and when Kumlien wrote to his "Friend Rasmus" on March 30, he said, "[It] has seemed to me confounded empty since you left & somehow it seems as though I made not as much use of the chance I had to see you, to talk & enjoy your presence as I should have done—you see I was sick when you came—but your presence acted as a good medicine and I have been better ever since." He also thanked Anderson for the gift of paint brushes for Ludwig, which put "new life in him & his pictures." Kumlien didn't know how Ludwig would do in oil painting, but he wanted Ludwig to have a chance to try it. But he also wanted the opinion of an artist on Ludwig's talent.

Then Kumlien went back to the idea of a position at the University of Wisconsin—which he had resolved not to pursue or think about. He said he had hope, faint and feeble, that he could be engaged at the university when he thought of how "exactly, precisely, entirely" such a situation would suit him, how he could "yet have some good of life" and keep his children in good schools, how he would have "easier opportunities for church going" for himself and his family.[40]

Finally, he suggests to Anderson that with his "ingenuity & sharpness, by setting in motion in that direction some of that large & fertile brain of yours, you may raise a breeze in my favor strong enough to cause immediate action."[41]

As Kumlien wrote to Anderson that March afternoon, asking him to help Kumlien get on at the university, Kumlien received a visitor. This man, known as Stiffneck, came bearing two letters: one from Anderson and one from John A. Johnson with an offer for a position for Ludwig. Johnson was an influential Norwegian American and Madisonian, a partner in a farm implement manufacturing company and a fire insurance company, and co-owner of the *Skandinaven*, a Chicago-based Norwegian newspaper. This influential Norwegian was lobbying for the establishment of the University of Wisconsin's Scandinavian studies department—which came about in 1875. We don't have the letter and we don't know what position Johnson offered Ludwig, but whatever it was, it required a "presentable wardrobe," which Ludwig didn't have. And it would have required Ludwig to leave immediately.

Kumlien declined the offer. He wrote to Johnson through Anderson and said the family had "considered all sides of the important question and the conclusion is that at present it is impossible for us to let Ludwig go."[42] He said that his wife saw it the same way, and so did Ludwig himself. Their reasons for declining the offer had to do with their debt and their reliance on their tobacco crop, which was reliant on Ludwig's labor. Kumlien wrote that tobacco, even at five cents a pound, "will give a snug profit,"[43] especially after their tobacco drying sheds were built. The family had a plan, which relied on all able-bodied members. And part of the plan was to send Ludwig Kumlien to the University of Wisconsin on tobacco money.

Though American Indians had grown small amounts of tobacco in Wisconsin before the arrival of European immigrants, tobacco, reintroduced in 1844, was a cash crop in southern Wisconsin by 1849. Primarily Norwegian farmers in Dane and Jefferson Counties raised and sold tobacco for chewing tobacco, snoos, or snuff rather than smoking tobacco. Tobacco was often the crop that paid the farm mortgage or paid for dairy barns or for college for sons and daughters. It is likely that tobacco allowed the Kumliens to finally, in 1874, build the frame house that Christine had wanted for so long. In the early 1870s, much of the Kumlien family's efforts were spent on building tobacco drying sheds and raising the crop, especially after Theodore—a natural born farmer—was old enough to take on most of the work. Often called a young man's crop because the stooped labor was hard on arthritic joints, tobacco was commonly tended to by family rather than hired labor.

The tiny seeds were sprouted in an old wool sock on the back of the cook stove, then the sprouts were moved to a tobacco bed covered with cheesecloth. After a few weeks, the larger sprouts were pulled and planted by hand in a freshly plowed and prepared tobacco field, usually fewer than four acres. The field of tobacco then had to be hoed for weeds, and the tops of plants and suckers had to be removed. The plants were carefully watched for insect pests and mildew, and when ready, they were cut just above ground level, strung on lathes, and hung from tamarack poles in the tobacco shed to cure. The dry tobacco was taken down from the sheds and stripped from its stems in December, bundled into forty- to fifty-pound bundles, and delivered to a warehouse in Edgerton.[44] In 1872, the Kumlien family had constructed a tobacco drying shed, with much of the work done by seventeen-year-old Theodore. And Kumlien, in a scheme to sell tobacco directly to a Boston tobacconist, sent samples of their tobacco leaves to Brewer. But Brewer, reporting that the tobacco wasn't as fine as Kumlien had hoped, ended up donating the leaves to his natural history society.[45]

Though growing tobacco was important to the family in the early 1870s, Kumlien was still collecting birds, nests, and eggs—some for Louis Agassiz at the Museum of Comparative Zoology at Harvard. One warm day in the summer of 1872, Thure and nineteen-year-old Ludwig set out for Lake Koshkonong to collect some of the more than two hundred pairs of Forster's terns they knew were breeding in the marsh. The nests, made of heaps of dead reed stems, were clustered among the new growth of reeds. Though the father and son saw many birds and nests and eggs, still they had difficulty getting even two or three positively identified sets of birds, since they soon discovered that the birds in the marsh were not all Forster's terns with orange bills and legs, pale gray wings, and black and white undersides. Among them were the similar common terns and Arctic terns. Because the day was hot and the birds didn't have to sit on their nests to keep the eggs warm, the entire colony of hundreds of terns hovered noisily over the Kumliens.

But Thure and Ludwig waited. Only toward dark, as birds began to settle on nests, could the men identify and procure a few sets of Forster's terns and common terns. When Thure and Ludwig returned to the colony in the marsh a week later, most of the eggs had hatched, but even these birdmen could not tell one species of speckled nestling from another.[46]

Agassiz, then one of the most famous scientists in the country, knew of Kumlien through letters and specimens Brewer had shared with him. In 1872 and early 1873, Agassiz, impressed by Kumlien's collecting and knowledge, sent Kumlien lists of birds, eggs, and nests he wanted for his Museum of Comparative Zoology, more usually called the Agassiz Museum. W. D. Hoard from Fort Atkinson, a promoter of the dairy industry through his magazine, *Hoard's Dairyman*, read in the *Chicago Tribune* that Agassiz believed Thure Kumlien of Busseyville, Wisconsin, to be the greatest authority in the world on bird nests. Hoard, who lived about ten miles from Busseyville, was surprised he had never heard of Kumlien. He wrote in a newspaper account that he "set out at once to see so unusual a man, and found him plowing his field with a yoke of oxen."[47] Kumlien did know about bird nests—not only where and when to find them but also the identities of the plants with which they were made.

Angie Kumlien Main, who grew up at Koshkonong and later became an ornithologist herself, wrote eloquently about her grandfather and bird nests. He searched, she wrote,

the fallen logs for the whippoorwill and nearby depressions in the ground for their nests; the marshes for the sandhill crane, the trees that hung over the creek for the nests of the wood duck, holes in the trees for the woodpeckers, bluebird, white-breasted nuthatch, owls, and the rushes on the lake shore for the nests of the red-winged and yellow-headed blackbirds, marsh wrens, the snipe, curlews, and Wilson's Phalarope. . . . He waded in water up to his knees and pushed his boat ahead of him through the reeds and sedges looking for rails [and] bitterns. . . . He searched the tops of old muskrat houses for the nests of the tern and, worst of all, he climbed hand over hand to the top of tall oak trees, where there wasn't a branch for fifteen or twenty feet from the ground, for the nests and eggs of the hawks.[48]

On November 12, 1873, Kumlien made a list of the eggs, nests, and birds he sent to the Museum of Comparative Zoology. Because Agassiz died about five weeks later, what could have been a valuable connection for both men never really began.

In October and November of 1873, Brewer wrote eight letters to Kumlien "principally in reference to the living ducks."[49] In a somewhat hilarious series of letters, we learn that Brewer has hatched a scheme for Ludwig Kumlien, with his father's help, to send him live ducks. Kumlien must have written Brewer about Ludwig's abilities to raise ducks and other birds whose eggs he collected by putting them under setting hens. Ludwig also saved the lives of wounded ducks that he or his father had shot. In order to carry out "ornithological experiments," and to help Ludwig, Brewer asked the Kumliens to ship live ducks to him.

At first, Brewer asked for birds that were "not badly wounded or with open sores, so that there is a fair probability of their suffering no inconvenience upon their journey."[50] He requested that the canvasback, redhead, and mallard be sent to him in one large box and that the Kumliens "make arrangements" with the express man and write "bird" on the box. He also asked them to add a tame drake to the box to "associate with these three female ducks . . . and see what comes of it."[51]

Things didn't work out well with the first shipment. Something—a woodchuck, a skunk, or a stray dog, Brewer speculated—got one of the ducks. Next, Brewer proposed that if they sent "a redhead and all your canvasbacks and the nest," the Boston Natural History Society would pay young Ludwig "not less than forty if at least four of them get to us alive and not more than fifty at the most."[52] Brewer suggested that Ludwig hold off putting feed in the shipping crate because the ducks might eat it all at once. The express men could feed them, Brewer said, "but can they be counted on?" Not surprisingly, he noted, "my colleagues shake their head at me."

On November 22, Brewer reported that the supposedly tame drakes the Kumliens had sent had eaten the hay in the crate, and two of them were dead. He instructed Thure and Ludwig to wait before they sent more, but to send all the notes they could on water birds."[53] On November 29, he wrote, "Better send me another installment of the ducks."[54]

Ludwig kept a Virginia rail, a small waterbird, in an enclosure he built around a spring, exposing the rail to all manner of non-rail sounds—music from a flute, boyish shouts, and the noises of ambient marsh birds. The Kumliens noticed that "the Virginia rail possesses the power of ventriloquism to an extraordinary degree. . . . There was hardly any note or noise commonly heard in the marshes that he could not imitate, so wonderful were his powers of mimicry."[55] The family hatched eggs of the common moorhen, and the American coot, "setting the eggs under a hen and feeding the chicks on baked cake, composed of cornmeal, oatmeal, bran and beef."[56] They reared American bitterns, taking chicks from nests and keeping them through the winter. Ludwig raised a pair of American goldfinches and a pair of mourning doves, feeding them the way he had seen parent birds feed their young. Ludwig also tamed a great blue heron, which followed him around eating the frogs and fish he caught for it.

Since he was a boy, Thure Kumlien had loved and studied birds. But for many years, this love of his had been a lonely pursuit of birds as specimens. His son Ludwig, with his experiments and delights in live and tame birds, must have reminded Thure of the origins of his own passion for birds. Ludwig, well prepared by his father, had, at the age of sixteen, written a piece called "Ornithological Observations" recording the migrating birds he observed at Busseyville between March 9 and May 26, 1869. Ornithologist A. W. Schorger, who obtained the paper from Angie Kumlien Main, described it in 1944 as "probably the earliest formal list of migrating birds prepared in Wisconsin."[57]

Later in his life, Ludwig wrote that in April and May on Lake Koshkonong, they never tired of watching groups of American white pelicans swim up a bay to feed. Father and son must have stood quietly on the shore watching as the large white birds, all in a line, all with their heads and necks below water, slowly moved toward the shore, crowding the fish and filling the nets of their bills. When one caught a fish, the bird raised its head until the orange-yellow bill was nearly vertical and the fish, or most of it, was swallowed.[58]

Theodore, Thure and Christine's second son, seems to have been happy working on the farm and staying at home helping his mother. As an older man, he said to his daughter, Angie, "I can't remember when my mother wasn't in poor health, so I helped her with her heavy work. When I had time, I cleaned the floor, worked the garden, and did anything I could to make her work easier."[59] Angie recorded the family memory that, despite being often unwell, Christine "had welcomed many visitors who flocked to see her husband, had fed them, and often had kept them overnight."[60]

Christine Wallberg Kumlien— quiet center of the family, essential helpmeet, dearly loved wife and mother—is rarely mentioned by Thure in his journal or his letters. There is no explanation for this

This image was copied from a tintype of Christine Wallberg taken on September 3, 1863. WHI IMAGE ID 147543

other than, perhaps, a reticence on his part that was cultural and of the era. Also, knowledge of Thure and Christine comes to us primarily through his writing to naturalists, old friends from Sweden, and young American protégés—all male. Almost no women appear in the writings we have of Thure Kumlien's. No description of Christine survives. Nothing we know about the Kumlien family allows us to see her as anything other than a good wife and mother; a talented and hard-working housekeeper, gardener, and farm worker; and a woman friendly with and helpful to her neighbors.

Probably as a result of the labor of his sons and the incoming checks from their tobacco crops, Kumlien was finally able to plan for Christine a comfortable frame house on the road north of their log cabin in the summer of 1874. Christine would finally have a house much easier to keep clean and free of pests than a log cabin. "How she worked and planned for that new home!" her granddaughter wrote.[61] All summer, the family worked on the new house. At this time, though there are no surviving records, Kumlien was likely still teaching at Albion Academy; he was said to have held a position at Albion "into the 1870s."[62]

As the construction on the house got underway, it became clear that Christine was seriously ill with cancer of the stomach. She was confined to her bed in the log cabin. Sophia Wallberg, who made her living in other homes caring for the sick, came to Koshkonong to nurse her younger sister. Christine died on September 22, 1874, just a few weeks before the house was completed. She was fifty-four. The next morning, Kumlien asked his neighbor Thomas North Jr. to go to Edgerton for him and buy the coffin. Christine's funeral was held in the nearly finished house. She was buried in the oak savanna of Sweet Cemetery next to their daughter Agusta, who had died almost thirty years earlier.

Sophia gave up her work and moved in to care for the family. The young Kumliens were twenty-one, nineteen, seventeen, and fourteen when their mother died. They called Sophia "Moster," which was Swedish for aunt. She was more firm with them than Christine had been, and she saw to it that the boys did more "hustling and farming."[63] Sophia had lived on successful farms and knew that the Kumliens had to get more land under cultivation. Recounting what her father, Theodore Kumlien, told her, Angie Kumlien Main wrote, "At first the children resented what they called her 'bossing,' but afterwards they saw that that was what they needed—to learn how to work in order to help themselves" in later life.[64]

A few days before Christmas of 1874, the year Christine died, Kumlien wrote to his old friend Brewer, "I have been altogether slow in sending you the few remarks I have been making in regard to the ducks, but since my dear wife's death I have been very little like myself." He had even lost interest in his "so dear Natural History." The ties that bound Thure and Christine for thirty-one years had become stronger with age. Kumlien wrote Brewer that his good children seemed to take their mother's death more sensibly than he did. "Your very kind sympathy . . . does real good to my bleeding heart," he told Brewer. "The great difficulty is that it is so hard for me to reason with my heart."[65]

On January 7, 1875, Theodore Kumlien turned twenty and began to keep a journal in a pocket-size memoranda book he bought that day at Burdick's store for forty-five cents. At the time, Theodore was a boarding student in the winter session at Albion Academy, coming home as early as he could on Fridays and going back to school on Sunday evenings. For not quite a year, Theodore wrote short entries in the leather-bound notebook. Though he spent most of the week at school, he wrote few entries about other students, his classes, or schoolwork. Mostly, he wrote about the farm, his farm chores, and the family. And almost every entry began with the weather of that day. When he was at school, the one-line weather report was often the whole entry. Theodore was not a willing or happy student. Nor was he a naturalist, like his brother Ludwig. And though he could draw a bit, he was not an artist like Ludwig and Frithiof. What he seemed to want to do was farm, go to dances on Saturdays and church services on Sundays, and play baseball. It was Theodore's steady managerial work and labor on the farm that provided a financial cushion for the work of the naturalists and artists in the family. Though his journal almost never mentions Moster Sophia, her hard work in the house and direction of the farm work made possible the comparative ease in which the Kumliens at Koshkonong lived for the next decade. Theodore's 1875 journal gives us a detailed picture of the Kumliens' family life during this time.

Though Theodore's journal entries are terse, they clearly record his activities and how he felt about them. On Saturday, March 13, Theodore wrote, "I & John [Reson, a hired man] cut cordwood today. it comes rather tough. I started to study late this evening." The next day, Sunday, on a "raw disagreeable day," he noted, "Lud & Swea went over the lake visiting to Ira bingham's today. it has been awfully lonesome for me."[66]

On other March weekends, Theodore chopped wood, stripped tobacco, cut white oak fence posts, and cut stove wood. He didn't attend the spring session at school because there was so much farming to do, but he took Ludwig back to Albion despite the almost impassable muddy roads. By early April, in clear and warm weather, when the ground was drying fast and some farmers were plowing, the Kumliens hired a man for 75 cents a day to clear more acres for tobacco. Theodore plowed the tobacco fields, bought broadleaf tobacco seed, and sowed the tobacco beds. A hired man sowed oats by hand, with Frithiof dragging and rolling the seed in. Frithiof plowed. He and Theodore moved a rail fence. While Theodore was farming, "Dad and Feity [Frithiof]" and Ludwig went hunting. Sixteen-year-old Frithiof was used where he was needed. Sometimes he helped "Pa" collect and prepare specimens for the normal schools.

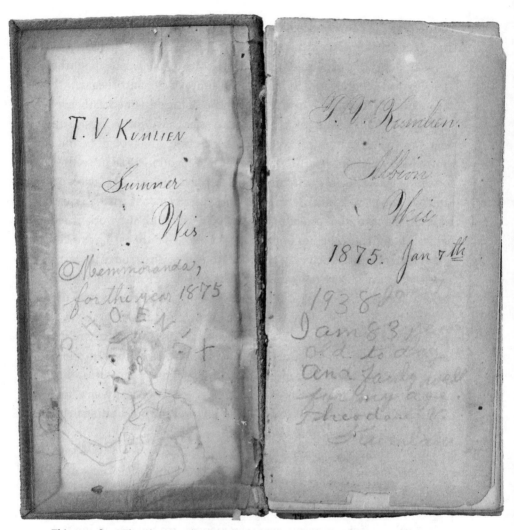

This page from Theodore Kumlien's journal includes a self-portrait in pencil. On the following page, he wrote his name and "Albion Wisc. 1875. Jan 7th" in ink. Below this, in pencil and in a more scrawling hand, he wrote "1938 Jan 7 I am 83 years old today and fairly well for my age." Theodore died three years later on March 28, 1941. THURE L. KUMLIEN PAPERS

Thure Kumlien had been hired by the first four normal schools in Wisconsin to prepare for each two hundred Wisconsin bird specimens to be used in the postsecondary preparation of the state's teachers. From 1874 to 1878, with the help of Ludwig and sometimes Frithiof, Thure collected and prepared more than eight hundred bird specimens for Whitewater,

Platteville, Oshkosh, and River Falls. Regent of Normal Schools J. H. Evans wrote from Platteville on January 30, 1877: "This is to certify that Prof. T. Kumlien had delivered in this school 200 specimens representing the Ornithology of Wisconsin. I further certify that the specimens are in excellent condition, artistically mounted and in all respects satisfactory to myself and to the faculty of the School."[67] This was an enormous undertaking. And it was, in a way, Thure Kumlien's "Birds of Wisconsin," more complete and certainly more accurate than Philo R. Hoy's 1851 list. Thure's son Ludwig would write the next one.

On April 19, 1875, Kumlien wrote Thomas Brewer for more details about Brewer's planned trip to natural history museums in Europe. Was he going to Sweden or Norway or Denmark? Was he going to do any collecting? Would he write to Kumlien from Europe? Brewer replied that the contents of Kumlien's letter "were

In this list of specimens Kumlien sent to Oshkosh, River Falls, Platteville, and Whitewater, he records the sex of each specimen and the date he either collected or prepared it. THURE L. KUMLIEN PAPERS, WIS MSS MQ, BOX 1 very pleasant reading." He said he very much wanted to go to Scandinavia, but might only get to Copenhagen, and he didn't think he would have time for collecting or letter writing. "P.S.," Brewer added, "I wrote you yesterday arranging the gift of a set of my book, from the publishers. I hope you will not object to writing such a letter of acknowledgment as they will want to print."[68] This book was the three-volume *History of North American Birds* by Baird, Brewer, and Ridgway, which contained at least nine references to the collections and observations of Thure Kumlien. In the first volume, Kumlien was either quoted or referred to in entries on the Wilson's thrush [veery], Cape May warbler, Philadelphia vireo, evening grosbeak, and lark sparrow. The authors described a lark sparrow nest sent to Brewer by Kumlien as "loosely intertwined stems of dry grasses, sedges, and carices" built on the ground, very flat with a shallow cavity.[69] In the second volume was a paragraph about the nest and eggs and parents of a "little fly-catcher" [willow flycatcher] collected on June 28, 1870, on the edge of Lake Koshkonong by "a son of Mr. Thure Kumlien." The outside of the nest, a large one for the species, was made of soft lichens and mosses and "within this a neat and firm nest woven of bits of wool and fine wirey stems of grasses."[70] In the third volume was a reference to a broad-winged hawk Kumlien collected in the vicinity of Lake Koshkonong.

These three illustrated volumes on American land birds were the culmination of Thomas Brewer's ornithological career. Spencer Fullerton Baird, the first commissioner of fish and fisheries for the US Fish Commission, became the second secretary of the Smithsonian Institution in 1878 when Joseph Henry died. Baird served as the Smithsonian's curator until his death in 1887. Robert Ridgway, taxonomist and ornithologist at the Smithsonian, contributed many of the illustrations in these three volumes. He later helped to found the American Ornithologists Union in 1883. When these three volumes on land birds were published, they represented the pinnacle of knowledge on American birds. It must have been a red-letter day when the volumes arrived at the Kumlien farm. Later, these three volumes would be the first bequest in Thure Kumlien's will. He somewhat surprisingly wanted his artist son Frithiof to have them rather than his ornithologist son Ludwig.

Brewer's April 1875 letter announcing the publication of his many years of work on the birds of North America was one of the last he sent to Kumlien.

During the previous November, Brewer had written that he was not in good health, though he was planning a trip to Europe. After his return from what must have been a taxing trip, he wrote few letters to Kumlien. The fruitful and rich correspondence between these two men was near its end.

That spring of 1875 was a busy one. Frithiof and Theodore hauled nine loads of manure to the "Burcander" field—Sven Bjorkander's name had been Americanized by the younger Kumliens to Burcander. While Thure and Ludwig worked on the birds for the normal schools, Theodore packed tobacco, sowed potatoes, and planted maple trees below the house, by the spring and by the old red gate. Theodore, and probably other Kumliens, went to Albion Academy to hear Ludwig give a talk on the golden eagle. And then one day when Frithiof and Theodore were mending a fence, someone from a circus showed up with a dead mountain ibex, a species of wild goat. The circus people either gave it to Thure or he bought it, perhaps thinking he could mount it and sell it, but in June, Ludwig and Swea took the prepared ibex to Janesville and then "fetched it back home."[71]

In the summer of 1875, Thure Kumlien and his family had many visitors. Naturalist Philo Hoy came from Racine one evening and stayed overnight. On May 20, world-famous Norwegian concert violinist Ole Bull spent the night. A virtuoso, Romantic Norwegian nationalist, and friend of Rasmus Anderson, Bull would have had a great deal to talk about with Kumlien. The next day, Kumlien took Bull to Edgerton to catch the train back to Madison. On July 29, Professor Rasmus Anderson came from Madison and spent a night. Carl Hammarquist came for a visit one day, and Gustaf Mellberg came to take the census. Theodore recorded in his journal, "Dolf Greene our old teacher called on us this pm on the 8 of July, the day the chinch bugs got the wheat." That year, the chinch bug destroyed most of the wheat and barley in the area. In July, all the Kumlien sons hoed tobacco and went haying on the marsh, hauling the cut hay to their hay mow. Thure prepared his birds. On August 8, seventeen people came to visit Thure. And the following day, ten more visitors came calling, drawn, it seems, to this educated Swede who could talk about all manner of things and who had wonderful birds to show them.

Though an August frost killed part of their tobacco crop, Theodore, who was nearly finished with the tobacco drying shed he was building, went to a Norwegian dance and was gone almost two days. On another day, he and Swea went to a dance in Busseyville and got home at daylight. Theodore also played left field—in Bunting's pasture—for the newly organized Busseyville baseball club.

On September 8, 1875, Ludwig went off to begin his studies at the university at Madison. His father drove him to the depot at Edgerton. The problem of money for Ludwig's education was likely solved with the help of tobacco money, which came largely from Theodore's planning and labor. Ludwig would attend Madison for most of four years, but his schooling would be interrupted by a chance to work at the US Fish Commission in Michigan, a wonderful opportunity for a young naturalist.

That September, Theodore happily stayed home suckering and cutting down tobacco, hauling loads to a barn, and ringing and hanging tobacco. He was building another tobacco shed, hauling lathes and shingles, buying nails, hiring a man to help. Because Theodore was so competent, Thure was able to spend the month working on the collections for the normal schools, and Ludwig was able to go to the university. On September 22, Theodore wrote, "One year ago today Ma died."[72]

It may have been that fall when Thure Kumlien invited William D. Hoard of Fort Atkinson to come to Lake Koshkonong with him to watch the migration of the canvasback ducks. Because Lake Koshkonong was by then a duck hunting destination for hunters from Chicago and Milwaukee and all over the Midwest, Hoard would have known about the canvasbacks. But he wouldn't have known what Thure knew. Hoard wrote that Kumlien took him to the lake before daylight, where the two of them lay flat on their backs in boats at the shore waiting with their faces to the predawn sky. And they did not have to wait long before they heard a roaring in the distance, a roaring of thousands of wings. "As they flew over us," Hoard wrote, "the noise became greater. On and on they came, great hosts of them." When the sun came up, Hoard saw that the water was covered with the beautiful ducks as far as the eye could see.[73]

In the fall and winter of 1875, Theodore attended one session at Albion Academy but never went back to school after that. Instead, he farmed the Kumlien place—a life of repetition and routine, though it was sometimes broken up by notable events. On December 3, he wrote, "I and Pa went to Stoughton this morning. Pa had some photographs taken."[74] They were delivering birds to the normal school at Stoughton. And one Sunday that

Thure Kumlien in 1875. WHI IMAGE ID 147678

winter, Theodore went to Sunday school and brought the neighbor boy Tommy North home to dinner. Many years later, Thomas "Tommy" North's stories about Thure Kumlien would inspire wonderful books by Thomas's son, the writer Sterling North.[75]

The end of 1875 was a time of reflection for Thure Kumlien. He often walked the path through the Bjorkander fields back to the log cabin in the woods where he and Christine had lived for so long. When he had time to spare, he liked to spend it in their old house where he could be alone with his thoughts and work on preparing birds.

One October afternoon in the old cabin, after he had stuffed a golden-crowned kinglet, Kumlien sat in the warm sun with a pencil and an old envelope and wrote a poetic meditation on his life, using the cabin as a literary device.

Kumlien describes the cabin as pleasant company, a company "eloquent and true." It was an old friend, his "dear old log house," and when he looked at it, he couldn't "help but feel a little strange." "Yes it whispers," he wrote, "yes it talks, yes and it speaks aloud to me of old times." Then Kumlien addressed the cabin directly: "Our present isn't much and our future prospects still less, my timbers are partly gone up, so are yours—age is upon me—so with you. With a little tender care I may last and be good for something yet a little while—so may you." Neither he nor the cabin were "cut out for pretensions and show in the world." Circumstances, he wrote, "put me in a kind of out of the way place not very conspicuous to the public, yet many are they who have visited me. So with you." Alluding to his life in the cabin with Christine and to her death, Kumlien wrote: "At the same moment we both lost our best friend, one who did more for us both than anyone else ever did." Finally, he summed up his life: "I have after all, been a comfort to some—perhaps you have too. I have served the purpose for which I was made. Have you?" After all the uncertainties and difficulties, the dissatisfactions and the sadnesses, of his earlier life in this cabin, the fifty-six-year-old Kumlien felt he had come to a place of calm and peace.[76]

13

RISING SONS AND DISAPPEARANCES

1876–1880

By the late 1870s, Thure Kumlien was no longer an obscure citizen "shut up in the woods," as he once described himself. In earlier years, he had inhabited two worlds: one being the small, close world of family and Koshkonong neighbors and the other being a larger, though distant, world of birdmen in the East and in Europe. In the 1870s, Kumlien also began to live comfortably in a middle world as a teacher and writer. He was becoming a citizen not only of the Koshkonong marshes and swamps but also of Wisconsin, where for the first time he would become a tourist. For years, his natural shyness, hesitancy to speak English, and poverty had kept him from socializing the way he had as a young man in Sweden. But now many forces were drawing Kumlien into the world.

One of these forces was Kumlien's American sons. Like many children of immigrants, they were a kind of vanguard, showing their father how to navigate the world around him. Frithiof's developing talents as a musician and artist made him popular among neighbors and friends. Ludwig, who was becoming known as a gifted naturalist, traveled back and forth between the university in Madison and the US Fish Commission in Michigan. And Theodore's handling of the farm not only freed Kumlien from work he found onerous but also freed him from one of the constants of rural life—the too-close scrutiny of his agricultural accomplishments by the neighbors. Kumlien's life simply became richer as his sons' comings and goings stitched him to their worlds.

The South Hall of Albion Academy, pictured here, was renamed Kumlien Hall in honor of Thure Kumlien in 1876. The last academy building standing in 1959, it was conveyed to the Albion Academy Historical Society to become a museum. It burned to the ground in 1965. A two-story replica, named Kumlien Hall, was later rebuilt and today houses the Albion Academy Museum. WHI IMAGE ID 28135

After these years, and for the rest of his life, Kumlien would be sought out in Wisconsin for his knowledge, for his memory, and for his very pleasant company. In the 1870s, he would become connected to Wisconsin beyond Koshkonong not only through his sons but also through his writing; through his membership in the Wisconsin Academy of Sciences, Arts and Letters; and through his teaching. Fifty-seven years old in 1876, Kumlien seemed to think of himself as an old man, yet his most rewarding years were ahead of him.

Kumlien had resigned from Albion Academy in 1869 in protest of the treatment of Rasmus Anderson, but by 1872 and perhaps earlier, he was back on the Albion faculty teaching botany, ornithology, and zoology and curating the academy's museum. And the students and faculty at Albion honored Kumlien in a way usually associated with honoring the wealthy or the dead.

On December 19, 1876, some students and faculty of Albion held a meeting in the room over Burdick's store to form an organization they called the Kumlien Society. The first regular meeting of the Kumlien Society was held the next month in what had been called South Hall but had just that year been renamed Kumlien Hall. That building burned in the 1960s, but a new museum was built in its place and is called Kumlien Hall. Though one might assume the Kumlien Society was a natural science society, it was formed to present learned academic lectures to the public. No one else at Albion was quite so exceptional or so honored.

In addition to his science classes, Kumlien taught Latin for a short time at Albion. Though Latin was a dead language, Kumlien was said to speak it readily. He believed that the way to learn Latin was to speak it, "to have it at the end of the tongue, so to speak."[1] So, rather than asking his students to ploddingly translate Cicero line by line, he spoke Latin in conversation with them. One student remembered that this method was ultimately "too slow for hustling America and Mr. Kumlien went back to his birds and his science."[2]

Another student noted, "Against other teachers there came periods of rebellion, moments of anger, but no harsh desire, no unkind wish ever rippled the current of good will between Professor Kumlien and those he taught."[3] When students remembered Kumlien, they universally mentioned his kindness and gentleness. However, it was his impressive knowledge that prompted the school to name buildings and societies after him. Kumlien's knowledge was especially visible and memorable when students visited the long log building where he prepared and mounted birds, where the sides of the room were hung with birds and mammals awaiting his attention.[4] As they wandered from American white pelican to whooping crane to bald eagle to tiny warbler, many of these young people had their first experience of seeing up close the wild creatures of their natural world and of being guided by one with such a depth of knowledge of that world.

Dr. Philo Hoy, the recently elected president of the Wisconsin Academy of Sciences, Arts and Letters, arranged to visit Kumlien at his farm on May 18, 1875. Ludwig Kumlien met Hoy at the depot in Edgerton and brought him to the Kumlien farm where he stayed a night, as Theodore mentioned in his journal. It may have been during this visit that Hoy recruited Kumlien to join the Wisconsin Academy, to be a natural history counselor for the academy, and to write an essay on the plants of Koshkonong.[5] If this is the case, we can be very grateful to Philo Hoy.

Visible in this photograph of the Kumliens' front room is a shelf in the right corner, likely made by Thure after he took a job at a furniture mill in Jefferson in the 1850s, and, among other stuffed specimens, one long-tailed weasel. WHI IMAGE ID 147688

The Wisconsin Academy of Sciences, Arts and Letters, which was founded in 1870 "to gather, share, and act upon knowledge for the betterment of Wisconsin,"[6] met several times a year, often in Madison, where members read papers on diverse subjects but most often on science. Thure Kumlien's name first appears on membership rolls in 1875. When Kumlien joined, founders John Wesley Hoyt, Increase A. Lapham, and Thomas Chamberlin were still active, though Lapham died near the end of the first year of Kumlien's membership.

Transactions of the Wisconsin Academy of Sciences, Arts and Letters—the journal of the Wisconsin Academy, now called *Wisconsin People & Ideas*—was and remains an important window onto the finest of Wisconsin thought in science, literature, and the arts. Of course, Kumlien, a scientist and lover of literature and the arts, was attracted to this organization and to the journal. In the third volume of *Transactions*, published in 1876, are two of the last three papers written by Increase Lapham, "Oconomowoc and Other Small Lakes of Wisconsin" and "Embryonic Development the Same in Plants as in Animals"[7]; papers on fish culture, Wisconsin fisheries, and moths by Dr. Philo R. Hoy; and "Notes on the Geology of Northern Wisconsin" by E. T. Sweet. There were papers on early copper tools of Wisconsin, the ancient civilizations of America, kerosene oil, comparative grammar, and industrial education. In this issue, Kumlien was listed as an annual member and as one of the three counselors in the Department of the Natural Sciences of the Academy.

Also in this third issue of *Transactions* was the paper "On the Rapid Disappearance of Wisconsin Wildflowers; A Contrast of the Present Time with Thirty Years Ago" by Thure Kumlien, Professor at Albion Academy. Though the paper is listed as one of the papers read before the Department of Natural Sciences of the Academy, it is not clear if Kumlien himself read his paper at one of the sessions. The two-page paper is scientific and elegiac, understated, quiet, and beautiful.[8] Kumlien rarely wrote for publication, and this writing reveals much about his life and times and the world he observed.

First, Kumlien succinctly establishes where he is writing from, what he is writing about, and what his credentials are: "For the last thirty-two years I have resided in the vicinity of Lake Koshkonong, in Jefferson county, Wisconsin, and have during that time paid some attention to the Fauna and Flora of that locality, and have collected somewhat extensively in nearly all the branches of Natural History, particularly Ornithology and Botany."[9]

After that simple introduction, Kumlien's confident and authoritative voice emerges in this complex sentence with a clear narrative: "When I first came here in 1843, a young and enthusiastic naturalist, fresh from the university at Upsala, Sweden, the great abundance of wild plants, most of them new to me, made a deep impression on my mind, but during these thirty-two years a large number of our plants have gradually become rare and some even completely eradicated."

Gracefully delineating a long sweep of time, Kumlien continues:

When first I visited the place where I now live, the grass in the adjoining low-lands was five and six feet high, and now in the same locality, the ground is nearly bare, having only a thin sprinkling of June grass, Juncus tenuis and J. bufonius, Cyperus Castneus, here and there a thistle or a patch of mullein and in the lowest . . . parts some Carices [sedges]. As the land gradually became settled, each settler fencing in his field and his stock increased, some plants became less common, and some few rare ones disappeared; Lupinus perenis [wild lupine], among the first. But when all the land was taken up by actual settlers, and each one fenced in all his land and used it as fields or as pastures for as many cattle, horses, sheep, and hogs as could live on it without actual starvation, botanizing in this vicinity became comparatively poor.

Kumlien's tone becomes angrier as he mentions settlers who not only use every patch of their land for farming but put on it more animals than is good for the land or the animals. But, seemingly out of energy and breath, he falls back to understatement at the end of the paragraph, sighing that "botanizing . . . became comparatively poor."

"In the oak openings," he continues, "besides grasses of several species there were an abundance of other plants of which I will mention only some Orchids from a small piece of opening-land near my residence: snake-mouth orchid, downy rattlesnake plantain, small-flowered coral root, the Adam and Eve orchid or putty root, the lily-leaved twayblade, and the showy orchid, . . . of these only one or two can be found in the same locality now." Most of us living today have never seen these plants in bloom in the wild as Thure did. So many plants had disappeared that he does not try to mention them all.

"In the thick timber along the Koshkonong Creek," he goes on, "there is now but one lot of about 40 acres where the plants can yet be found nearly as abundant as formerly." In this unplowed spot along the creek, he says he can find the wild blue phlox named after Increase Allen Lapham, *Phlox divaricata, Laphamii*, as well as edible wild leeks or ramps; wide swaths of the down-looking, speckle-leaved trout lily; the cutleaf toothwart, also called crow's toes; Canada wild ginger; and many other rare and interesting plants.

Kumlien speaks of the local plant communities as if they are under siege, including the tamarack marsh he visited with young Edward Greene many years earlier: "A Tamarack marsh held out the longest; it was not visited by cattle till, for want of pasture elsewhere they were obliged to cross its miry borders. In this marsh, or on its borders were formerly growing: green adder's mouth orchid, fen orchid, prairie fringed orchid, green fringed orchid, lesser roundleaved orchid, dragon's mouth orchid, grass pink, yellow lady's slipper, small white lady's slipper, false asphodel, slenderleaf sundew, Kalm's lobelia [named after Kumlien's countryman Pehr Kalm, who came to America in the mid-1700s], . . . adder's tongue, white beak-sedge; Rannoch-rush or pod grass; the aromatic marsh arrowgrass." In both scientific and common names, this list of rare or disappeared plants tolls like the names of the dead.

"Now," he writes, "almost all of these and many other interesting plants growing in the marsh or near it . . . have become very rare and some are totally eradicated."

Only on land where plowing is impossible are some native plants still found: "On a small prairie, too stoney and gravelly for cultivation, there can yet be found prairie smoke, the rare annual salt marsh aster which grows on gravel bars, the half parasitic downy Indian paintbrush, a blue flax, the downy gentian, hairy hawkweed, Carolina whitlow grass, lyreleaved rock cress, and two Arenaria (stricta, Mich. and Diplopappus) which on gravel hills grows only two or three inches high, with leaves very stiff and narrow, but the flower large, having somewhat the aspect of an Alpine plant."

Kumlien compares the number of native plants he saw in 1876 to the number he saw in 1843: "A list of the plants of this vicinity, giving the plants of today, would be a comparatively meagre one and nearly useless, as their number is lessening every year, and a list of the plants of thirty years ago would have perhaps no other than a small historical value." He seems to lose heart in this sentence, stumbling a bit, as he contrasts in his mind the plants he found in 1843 with those he could find thirty years later, "when every day more and more of them disappear."

The next line is the last: "These observations, though made only in this locality, do probably apply to all the settled portions of the state." It's a quiet statement, informed and grim, a statement of grief Kumlien must make because he has witnessed this disappearance with his own eyes as few, if any, others have. He believes it is true not only of Koshkonong but

Corydalis solida painted by Thure Kumlien. WHI IMAGE ID 147673

also of all the other settled parts of Wisconsin. Settlement, Kumlien is saying, has all but destroyed Wisconsin's wild flora.

Kumlien's look back at the flora of the 1840s, written and published in 1876, has some of the same elegiac tone as his unpublished meditation on his log cabin. Both of these are successful pieces of writing—controlled, informative, and emotionally resonant. Yet one is personal, an extended metaphor that speaks of who he was, how he lived, and how he grieved for his Christine. These two published pages on the disappearance of Wisconsin's flora are Kumlien's most important piece of writing. His feeling, his voice, and his perceptions are there, but this is not simply self-expression.

It's an early observation of what was happening to the state—a masterful description of what was here and what was lost. Thure Kumlien and many others from Sweden, New York, Pennsylvania, England, Ireland, and Germany had dreamed of cabins in the woods, little places where they could live freely and make a living on a small farm. But if we see, as Kumlien did, the detail of what each dream meant to forty acres of a native landscape, and if we multiply that one dream by hundreds and thousands, we can begin to see some of the costs of these dreams.

After reading this paper, recognizing Kumlien's writing ability and knowing of his exceptional knowledge of Wisconsin fauna, Edward Asahel Birge, a professor of natural history at the University of Wisconsin at Madison, spoke to Kumlien about the piece, likely at a Wisconsin Academy meeting. Birge also wrote to Kumlien several times, urging him to give the academy another such paper, one on mammals or birds. "You know a great deal about the habits, etc., of our birds and mammals which no one else knows and much of which no one else can know now," wrote Birge on September 17, 1879. "It ought to be put on record and I hope that you will find time to set down some of it before next Christmas." Birge wrote again in 1882 urging Kumlien to write what he knew.[10] Kumlien would write for publication once more, but not exactly as Birge had hoped.

In June 1876, Charles Mann, a scion of a large Wisconsin manufacturing family, wrote Kumlien to thank him for two boxes of live plants, including white lady's slipper, which Mann had seen bloom for the first time in Philo Hoy's Racine garden a few days earlier. With warehouses and factories in Chicago and Milwaukee and two large factories in Two Rivers, the several generations of the Mann family were powerful merchants and manufacturers. Though we don't know how Mann met or connected with Kumlien, it is known that Mann fancied himself a naturalist and planned to grow all the ferns of Wisconsin in his Milwaukee garden. From hunting trips in Colorado, he brought back skins of wolverines and other mammals, which he asked Kumlien to mount for him. Mann also knew Kumlien's oldest son and asserted in a letter to Kumlien that "Ludwig would make his mark."[11]

Ludwig Kumlien was already beginning to make a mark. Even at the young age of twenty-two, he was known by many to be a reliable and talented observer of the natural world. Some of his ability may have resided in what his future wife would describe as "phenomenal eye-sight [that] allowed nothing to escape his observation."[12] Fluent in both the drawing and writing of his observations, Ludwig was well trained, of course, by his father. In 1875 and 1876, Ludwig traveled to the East and West Coasts as part of his job for the five-year-old US Fish Commission, headed by the all-around scientist Spencer Fullerton Baird with whom Thure had corresponded about birds. The Fish Commission was founded to study the causes of the decrease of commercial fish in the coastal and inland waters, but it soon counted fish in waters all around the country, raised fish, started canneries, and moved fish roe and fingerlings around the country to plant them in distant waters. In an 1877 letter to his son Theodore, Kumlien wrote that Ludwig was "to assist in distributing salmon in the western rivers."[13] Ludwig may have also tended fish tanks in baggage cars as the fish were carried across the continent. Between trips for the Fish Commission, Ludwig attended the university at Madison.[14]

On July 24, 1877, Thure Kumlien took the train to Milwaukee and then, with a free pass as a naturalist, took a steamship ninety miles up Lake Michigan to Two Rivers. Kumlien was on a rare pleasure trip—he was briefly a tourist. On his two days in Two Rivers, Kumlien was impressed by the large furniture and woodenware factories of the Mann Brothers. And he went botanizing on the dunes "in the strange pine region," collecting some plants he had never seen before and "some old country plants" that he hadn't seen in thirty-five years, but no birds.[15]

In Milwaukee, many things were a "wonder" to him. He wrote to Theodore, who was off in Iowa working for a harvest season, about the great waterworks, the state fish hatcheries, and the Soldiers Home, one of only two in the United States, "where 750 soldiers are kept and provided for with comfort and plenty." He toured the Forest Home Cemetery, calling it "one of the most beautiful cemeteries in the US with parks and artificial lakes and miles of beautiful flower beds."[16]

Kumlien wrote that he had been gone just a week, but it seemed like a long time even though he had nothing to do but try and enjoy himself. Everybody was kind and overattentive to him. Fond father that he was, he said the trip would have been more enjoyable if his children were with him to share the sights. And though he thought Milwaukee "such a nice place," he said he wouldn't want to live there half as much as at his "dear old home."

From the train window on his way home, he saw the fields of ripening oats and corn and wheat and thought that "our neighborhood is as good a country" as any he saw.[17]

When Kumlien got home, everything seemed to have grown a foot or two, "even the pigs have grown and the hogs and all."[18] And a Janesville man had left him forty South American bird skins to prepare, birds "more beautiful than imagination can paint them," he wrote Theodore. He would be busy that fall collecting and preparing bird skins and mounts. From the Frederick Kaempfer Company on Clark Street in Chicago, he could now order glass eyes for stuffed birds at five to fifty cents a pair, insect pins for thirty cents a hundred, and arsenic for twenty-five cents a pound.

In Kumlien's long, detailed, and happy August 3, 1877, letter to Theodore about his trip to Milwaukee and Two Rivers, he also mentioned that Ludwig would soon sail from New London, Connecticut, on a year-long exploration of the coasts of Greenland. But Kumlien didn't mention that he had just mailed off to Rasmus Anderson what would be the second of his two published pieces of writing.

Three weeks earlier, Kumlien had written to Anderson in Madison: "You know well enough that I am no writer, and when you got me to promise to write that thing—I felt sore enough. However when I must, I must, as the old woman said, when she had to swear, and here it is in a fashion." Because he thought the piece "wasn't much in the scientific line" and because it was as a scientist that he wanted to be recognized, Kumlien asked Anderson to have the piece signed "an old settler."[19]

Kumlien's second published piece of writing, "Lake Koshkonong," was published in 1877 in *History of Madison and Dane County*, edited by Madison bookseller and publisher William J. Park. The book is a compilation, both a guide and a county history, with pieces by writers in each town and city in the county. Albion, for example, is described by Dr. A. R. Cornwall. As he had requested, Kumlien's "Lake Koshkonong" in the six-hundred-page tome is credited to "an old settler." However, in the contents it is credited to "Prof. T. L. Kumlien," perhaps to the amusement of Rasmus Anderson. Or possibly at his behest.

Though Kumlien lived in Jefferson County, just a mile from the Dane County line, he was known to be an authority on Lake Koshkonong, which

was partly in Dane County. He had been asked to describe Lake Koshkonong, as it was in 1877, with perhaps some history, and at first, he does just
that. He writes objective descriptions of the lake—where it is in Jefferson
and Dane Counties, its orientation and shape, its points and bays, and its
shallowness, as well as the Rock River, creeks, and springs that feed it.
He describes Blackhawk Island, which isn't really an island, and the land
surrounding the lake—the marshes and limestone bluffs. Almost any
observer could have written this first page of the piece.

Then Kumlien writes,

> The lake, with its, in many places, marshy shore and hundreds of acres
> of wild rice, and the grass-like plant, known to botanists as *Vallisneria
> spiralis*, growing in it in the greatest abundance, used to be a great
> favorite place for ducks, and especially the far-famed Canvasback
> (*Aythya vallisneria*), which, with the Redhead, is particularly fond
> of the *Vallisneria spiralis*. Geese, cormorants [double-crested cormo-
> rant] and white pelicans [American white pelicans] were also very
> numerous, and fifty to one hundred of those latter birds could be
> seen at one time in the latter part of April or first of May.[20]

What "used to be" becomes the subject of the rest of this paper. The old
settler is telling what he saw and what is no longer there to see. And this
time he is writing about birds: "In the marshes and on the shore were a
great variety of waders . . . the great blue heron, the large white heron
[great egret], the snowy heron [snowy egret], the night heron [black-
crowned night-heron], and the least heron [least bittern]." That list alone
is heartbreaking, but he goes on. There were in abundance "six species of
the plover family and Wilson's *Phalarope*, the most beautiful of all our
waders. . . . Of the snipe family, twenty species, besides curlews and godwits. Three species of rails, and gallinules [common moorhens] and coats
[coots], very plenty."

Then Kumlien explains what happened to the birds: "But owing to a
continued sporting kept up every spring and fall for years, the birds have
either greatly diminished in number or found other places where they are
less disturbed, as now-a-days but few visit the lake compared with what
they did only ten years ago." After the coming of the railroads in the late

A Kumlien drawing of a duck among sedges. WHI IMGE ID 147671

1850s, a great number of unregulated sportsmen had arrived on the lake, now accessible from Milwaukee and Chicago. And in addition to the sportsmen were market hunters. "Ducks, even such as shelldrakes, whistlers and butterballs, bring something in the markets of the large cities," Kumlien writes, "and hence they must be killed and sold for the little they bring."[21] Kumlien had killed some of these birds, too, but one man with one shotgun collecting for knowledge could not do anywhere near the damage done by hordes of market hunters and sportsmen who would kill any bird, as Kumlien wrote, "for the fun of shooting them."[22]

"As for the fish in the lake," he continues, "the time is past when twenty-eight to thirty-five pound pickerels can be found, or twenty-five pound catfish. Bullheads and perch, sunfish, garpikes and dogfish are common yet; but the pike, pickerel, bass, redhorse, sucker and catfish are not near as plentiful as formerly."[23] For this, he blames the dams on the Rock River below the lake.

Kumlien then writes that because he is more interested in ornithology and botany than in archaeology, he can't give as much information as he would like about Indian artifacts. However, he will give a few facts that have come under his observation, mostly on the west side of the lake where he has lived for almost thirty-four years, "facts proving that this lake and vicinity have been a great resort of the Indians."[24] Though Kumlien must have observed these Indian relics for years, this is the first time he mentions them in writing that has survived: "On the land of Mr. R. Bingham are patches of ground where yet can be seen what is supposed to be cornhills worked by the Indians. While plowing or hoeing, Indian arrows, stone implements and pieces of pottery are frequently found; these relics are especially numerous on the farms of Mr. R. Bingham and Mr. Charles Lee, who has an extensive and interesting collection, picked up on his farms."

In addition, he writes, "Indian mounds of different sizes and shapes were numerous on the west side of the lake, but many of them are now leveled by the plows." He does not mention the more than twenty Indian mounds on his own property, formerly Bjorkander's field, which have not been leveled by the plow.[25]

The essay closes with Kumlien's mention of three remembrances. First, he describes a scene near his home: "At Busseyville, near the creek [Koshkonong Creek], there used to grow a very large oak which, thirty-four years ago, and at that time considered old, had a very plain and good figure of a mud turtle cut on the side, towards the creek, and on the hill north of it, were several mounds, some of which had the shape of mud turtles. These mounds are now leveled, and the land cultivated."[26]

Next, Kumlien recalls that thirty years ago, when he was botanizing near the lake, he found wild tobacco, *Nicotiana rustica lin*, growing in the grass on Bingham's land—where American Indians had lived. "This was at a time," he notes, "when the first settlers never had heard of raising tobacco in the state. Since which there is scarcely a farmer for miles around that is not engaged in raising tobacco."[27] Kumlien had stored away both of these early observations of Native people's lives, apparently without written comment, for more than three decades. We can't help but wonder what else he observed of the Native people who lived there before him, what else he had seen but never written about. Perhaps in these two memories, Kumlien is identifying with the way the Indians had used the land—changing it subtly but using what the land had to offer without destroying it.

Finally, Kumlien tells a story about a steamboat on Lake Koshkonong. He had suggested in his cover letter that Anderson might take out these last paragraphs, but Anderson and the editor left them in. Though a simple story, it reveals Kumlien's distance from and attitudes toward the area's settlers, both then and now. He writes:

> In 1844 there was a steamboat going through the lake, said to have come up from St. Louis. The new settlers hailed this occurrence with great pleasure and hopes, expecting to have a communication by water opened with the cities on the Mississippi river, and having no railroad nearer than Buffalo, N.Y., and sixty to seventy miles to haul their grain to Milwaukee, it is no wonder that they considered the coming up of this steamboat as a Godsend. The idea never occurred to them that this big Rock river, on which with their own eyes they had seen a steamboat from St. Louis, ever could be, by any authority, pronounced an unnavigable stream, and dams allowed to be built across it.[28]

Kumlien does not count himself among the settlers; he refers to the settlers as "they," not "we." As a well-educated Swedish naturalist and recent immigrant to the Wisconsin Territory in the 1840s, he had been an outsider among the single-minded settlers. As he wrote this essay in the 1870s, his disdain for what he saw as the settlers' foolishness also distanced him from them. Kumlien was still, to some extent, an outsider.

The piece concludes with the description of a "steamboat on the lake now," a boat built on the lake, a boat that could never leave the lake because of the dams, circling "between Taylor's point, Fort Atkinson and Newville," an image Kumlien finds pathetic and vaguely comical.

Though Kumlien wrote only two pieces for publication—this one for *History of Madison and Dane County* and the other for *Transactions*—both reveal his superior skill in crafting English sentences as well as the range of his point of view and his voice—at times scientific, poetic, elegiac, sardonic, and critical. In these two pieces, it's evident that Kumlien knows he is writing, if not for the state or the ages, at least for a world beyond Koshkonong. In letters to Brewer, Kumlien's detailed observations, careful measurements, and scientific thinking led him to sound conclusions as a naturalist. These two papers provide short, tantalizing glimpses into Kumlien's impressive, subtle, and critical mind.

At the same time that Kumlien was working on the second of his two important pieces of writing, his son was approaching a significant turning point in his own career as a naturalist. Ludwig Kumlien, who had been working for Baird and the US Fish Commission around the Great Lakes and on the West Coast, had been hired earlier in 1877 to be the "representative of the Smithsonian Institution, under the instructions of Professor Spencer F. Baird, the distinguished naturalist, for the purpose of collecting specimens of the flora and fauna of the country" on the Howgate Preliminary Polar Expedition.[29] The 1877 first phase of this two-phase expedition would send Ludwig and meteorologist Ornay Taft Sherman on a whaler, the *Florence*, to Annanactook on the Cumberland Sound in Greenland. There, the two scientists were to establish relationships with the "Esquimaux" and carry out scientific experiments, while the men of the *Florence* carried on their usual whaling. The second phase of the expedition would take place in 1878 when Captain Henry W. Howgate and others in a second ship would meet up with the men of the *Florence* to establish a polar colony for further scientific research. At least, that was the plan.

On August 2, 1877, the *Florence* set sail from New London, Connecticut, heading for Greenland. Kumlien wrote to Theodore on August 3 that he had received a letter from Ludwig, who was just about to sail. By the time Kumlien got the letter, Ludwig was "out on the great Atlantic."[30] Kumlien expected that he might hear from Ludwig one more time in the next fourteen months. Though his letters include no further comment about Ludwig and the Howgate expedition, Kumlien must have had a great mix of emotions when he thought of his twenty-four-year-old son being gone for a year or more to the Arctic. He must have been at least a bit worried for Ludwig, with the distance and the dangers. Kumlien also must have been immensely proud, for he knew that Ludwig—with his observational and scientific skills, his knowledge of birds and plants, his abilities to write and draw—was certainly capable of this work. But he also must have envied his son. This Smithsonian-funded trip to the Arctic for a year or two was just the sort of collecting trip Kumlien had long dreamed of taking. Kumlien had been at work as a naturalist for more than forty years, but other than his trip to Gotland Island, he had never been paid to travel and had not seen much of the world since arriving at Koshkonong. He went on with his life at the farm while he waited more than a year to hear from Ludwig.

In August 1877, when Ludwig left on his trip, Theodore was working in Iowa and Frithiof was home at the farm. During his trip to Milwaukee, Kumlien had bought three good guitar strings for Frithiof, who played the guitar when he wasn't painting or working on the farm. Besides the guitar strings, he had brought from Milwaukee Alfred Mann, the thirteen-year-old son of the wealthy Mann family, to stay in Ludwig's room and learn how to do farm work. Alfred's father, Charles, had purchased some plant specimens from Kumlien the previous year. That August of 1877, Kumlien, Frithiof, and young Alfred Mann were invited on a picnic excursion on Lake Koshkonong; they took a boat to Black Hawk Island and back. And on August 20, Frithiof shot a northern harrier that was feeding on something on top of a muskrat house near Lake Koshkonong. That something was a black rail, the first instance of this little bird recorded in Wisconsin to that date. [31]

Then Theodore came home from Iowa and Frithiof left for his first year at the university at Madison. Kumlien wrote his youngest child right away: What subjects were hard for him? Did he need money? He should go look at the three Gilbert Stuart paintings in the art museum. He should keep a stiff upper lip and dig hard. He should be sure and ask for money if he needed it. He should bring his dirty clothes home in his satchel. Did he need his quilt yet? Kumlien told Frithiof about the hogs they'd sold and the repairs on the tobacco barn. Would he miss too many classes if he came home on the Friday noon train and went back on Monday evening? And when he came home, he shouldn't leave the old flute behind. It would be needed in the family trio with Frithiof on guitar, Theodore on accordion, and Thure on flute.

On Monday, the first day of October in 1877, twenty-two-year-old Theodore loaded the family's wagon with six hogs and delivered them to a buyer. On Tuesday, the wagon now free, Theodore and Thure loaded it up with birds—birds that Thure had mounted for the normal school at Whitewater. Some of the birds they delivered that day were the Lapland longspur, bufflehead duck, Wilson's phalarope, whooping crane, American white pelican, black-throated blue and Nashville warblers, golden eagle, pygmy owl[32], sandhill crane, black-crowned night-heron, American bittern, and passenger pigeon.

Kumlien is curiously silent on the passenger pigeon. He had certainly seen them and collected a few, calling them in the 1840s simply "pigeons." One mounted passenger pigeon was sent to each of the four normal schools for which Kumlien was collecting educational specimens.

Though Kumlien had seen flocks flying over, the birds' great roosting sites among the beech and white oak were farther north than Koshkonong. On three days in April and May 1869, Ludwig recorded three great flocks of thousands of passenger pigeons, one flock "flying south, very high and slow," another "flying east very high, early in the morning," and the third flock flying north. That year, Ludwig collected two passenger pigeon nests, one high in a small elm and the other high in a black oak.[33] Clearly, the great flocks did not nest there in summer—at least in 1869. But it is curious that nowhere in the writings of Thure Kumlien does he describe or respond to the elegant birds, larger than mourning doves, swift in flight, and traveling in flights of uncounted millions that filled the sky with strange sounds, broke limbs from forests of trees as they roosted, and covered the ground with white droppings like snow. Like most people of the time, Kumlien probably could not imagine that the most numerous bird in Wisconsin could one day become extinct. We do know that he and his son abhorred the "netters" of the passenger pigeons, as they abhorred the market hunters who descended on Lake Koshkonong's ducks every spring and fall after the early 1870s.

Even worse than netters were plume hunters, who systematically slaughtered birds for their feathers to decorate women's hats. Ludwig described a marvel he and his father had witnessed each spring until about 1875 when thousands of Bonaparte's gulls would flock together on Lake Koshkonong and then, in the same moment, "with one accord," begin their circling flight to the north. "These vast flocks passed directly over our house as they left the lake," Ludwig wrote, "and many a time have we watched them, rising higher and higher, and gradually fading from view."[34] Then, in the 1880s, two men from Chicago began regularly coming to Lake Koshkonong to shoot the Bonaparte's gull for its plumage. On other lakes, other plume hunters systematically slaughtered the Bonaparte's gull. Ludwig reported that this massacre had so reduced the gull's numbers that it was no longer the most abundant species of the area. In addition to the Bonaparte's gull, the formerly common American bittern was greatly reduced by "idiotic practice of so-called 'sportsmen.'"[35] The sportsmen killed the bitterns as they flushed them from the grass and the shores of lakes.[36]

On October 30, 1878, over one year since he had set sail in August 1877, Ludwig Kumlien arrived back in New London on the *Florence*. Instead of the more-than-two-year trip he had expected, he was returning after having been in the Arctic for just about fourteen months. When the *Florence* had arrived at the rendezvous point in Godhavn Harbor in July, the men had expected to meet up with Captain Howgate and the second ship, at which point they would set off to establish a polar colony. However, when the *Florence* arrived at the harbor, they got word that there was to be no second phase of the expedition after all. So, on August 22, the *Florence* had set sail for home.

By April 15, 1879, the report of the Howgate Expedition was published by the Smithsonian Institution, introduced by Baird as the fifteenth in a series of papers on the natural history and ethnology collections of the Smithsonian Institution. The report was titled *Contributions to the Natural History of Arctic America Made in Connection with the Howgate Polar Expedition, 1877–78*, and the author was Ludwig Kumlien. The first 106 pages of the 171-page report are taken up by four sections written by Ludwig: Introduction, Ethnology, Mammals, and Birds. Tarleton H. Bean, an ichthyologist at the Smithsonian, wrote more than thirty pages on fishes. The remainder of the book comprises lists of species collected by Ludwig and identified by experts in the fields of crustaceans, mollusks, insects, and

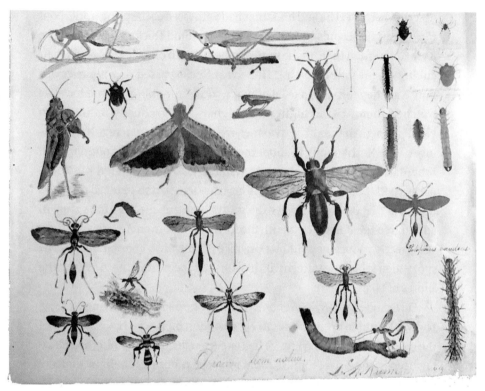

This page from Ludwig Kumlien's 1860s notebook includes a katydid, stink bugs, milli-pedes, wasps, and a violin-playing cricket. COURTESY OF NANCY KUMLIEN-GARRY

more. Plants were identified and catalogued by Asa Gray at Harvard and lichens by Edward Tuckerman, who had in 1841 visited Elias Fries at Uppsala. The report was written clearly, though quickly, in Washington where Ludwig stayed for a few months, working under the supervision of Baird. At the same time, Captain Tyson wrote his report, *The Cruise of the Florence; or, Extracts from the Journal of the Preliminary Arctic Expedition of 1877–'78.* This, too, was published in Washington in 1879.

In his report, Ludwig described the discouraging conditions under which he had to prepare specimens—in a cold and dark snow-hut or igloo on shore. Treacherous ice and stormy weather kept him from searching very far from the ship, which was frozen into the ice. The ink would freeze in Ludwig's pen as he took notes and sketched by the dim light of a seal oil light. At one point, he cut his finger while skinning a dog that had died of a virus, and he lost the use of his arm for several months. Though Ludwig did not complain in official publications, he later wrote that the captain had unnecessarily kept him on board the ship when he needed to be collecting in the field. However, no dissension about the expedition of the *Florence* to the Arctic was made public.

Several years later, after the army and navy had withdrawn support for the 1880 polar expedition, Howgate resigned from the army and found private funding to send the *Gulnare* to the Arctic. The cruise of the *Gulnare* was short-lived due to storm damage to the ship, and it returned to Newfoundland. Howgate, as he raised money for another expedition, was indicted in 1882 for embezzling the hundreds of thousands of dollars intended for this next expedition. After eluding capture for twelve years, Howgate was finally caught in New York City, where he had been living with his mistress and dealing in antique books. After that, he lived in the Albany Penitentiary.

Yet, this scandal did not taint the fine report of Ludwig Kumlien or his reputation as a naturalist, which he hoped to parlay into a full-time job in Boston, Washington, or New York. Though he had the support of Baird at the Smithsonian, the work Ludwig sought as either a naturalist or an illustrator eluded him. He returned to the Great Lakes area in 1879 and worked again for the US Fish Commission, gathering statistics on fisheries.[37]

We have no comment by Thure on the report Ludwig prepared for the Smithsonian. However, he must have been pleased when the gull that Ludwig found and reported as *Larus glaucescens*—now known as the glaucous-winged gull—was identified by William Brewster as a new species, which he named *Larus kumlieni*, or Kumlien's gull, after Ludwig. That gull (*Laurus glaucoides kumlieni*) is now considered to be a subspecies of *Larus glaucoides*, or the Iceland gull.

But another era closed when Thomas Mayo Brewer died in Boston on January 24, 1880. Though Kumlien and Brewer corresponded for almost thirty years and certainly became friends, they never met.

In the years leading up to his death, Brewer had become embroiled in what became known as the Sparrow War. As the introduced and invasive house sparrow, or English sparrow as it was then known, moved west across the continent, ornithologist Elliott Coues fiercely and publicly raged against this invasion, and he disagreed with Brewer, who defended the sparrows against those who would trap and poison them. Kumlien—who had been delighted just a few years earlier to see again the house sparrow, which he had last seen in Sweden—would have sided with Brewer against Coues's harsh attacks of Brewer and the sparrows. Though the younger Coues was correct about the damage the invasive sparrows would cause to native American birds, Brewer and Kumlien had lived long enough to understand that the sparrows were here to stay and that the natural world they remembered had already begun to disappear.

14

HOSTS OF BEAUTIFUL THINGS
IN THE WORLD

1880–1888

In 1880, Thure Kumlien was collecting for the Natural History Society in Milwaukee and for Spencer Fullerton Baird at the Smithsonian. In September 1880, he sent Baird a rare find, though one he may not have collected himself—a male trumpeter swan, including its "very curious & interesting windpipe arrangement."[1] Mounted to show its five-foot wingspan and its beak-to-tail measurement of sixty-five inches, the swan, which Kumlien had sold for twenty dollars, must have been shipped in a large and carefully rigged box.

Kumlien still lived in his cabin in the woods, with Sophia Wallberg keeping house. Sophia was in her early seventies, a small woman, "thin and peppy . . . with very fine straight black hair with a few gray hairs."[2] Every morning, she put on a big gingham work apron over her dark dress. Every afternoon and on Sundays, the apron was white. In the mornings, she roasted green coffee beans in a frying pan over the wood-burning cook stove and ground them in a grinder attached to the bottom of the stairs.

Theodore Kumlien had married Mary Alice Langworthy, who went by Alice, in Albion on March 14, 1879. After settling in on the farm, the two began raising their family in the frame house. By 1883, Thure was the grandfather of three: Margretta, called Madge; Charles Theodore, called Charlie; and Mary Angelina, called Angie. Angie would become an

Sophia Wallberg was said to be hard on noisy children. WHI IMAGE ID 147682

ornithologist and her grandfather's biographer.

Theodore had purchased the north forty acres of the farm, the Bjorkander forty, from his father for nine hundred dollars, though Ludwig objected. Ludwig wanted his father to sell the land to someone who would pay more and pay in cash because he had in mind that he and his father would use the money for a long trip to Sweden. Fair-minded Thure wrote Ludwig explaining that Theodore "will have hard work, harder than any of us, and he is my son and your brother, and when I am through there will be, if I steer straight and don't get much sick—about $600 for each of my children."[3]

After Swea Kumlien married Robert Stratton Martin in 1882, the couple moved to Cedar Mills, Minnesota, where he taught school. Swea was her Aunt Sophia's favorite, according to a family memoir. Sophia was closest to Swea and Frithiof, who were seventeen and fourteen when their mother died.

Frithiof, the youngest, graduated from the University of Wisconsin in 1882 with a bachelor of science degree in pharmacy, then moved to Milwaukee where he became a student at the Layton School of Art.

Ludwig had been chasing employment since he had completed his report on the Howgate expedition for the Smithsonian. He tried unsuccessfully to get work as a natural history illustrator at a museum and at a publishing house, but he was rejected by the publisher because his drawings were not "pictures enough."[4] Ludwig wrote applications and letters to his friends at eastern museums, but nothing worked out. When Ludwig couldn't get other work, he would return to the US Fish Commission for a season, but he didn't like the work. In 1881, Ludwig was in Milwaukee

According to notes on this photograph, the tree and vine pictured here mark the division of Theodore Kumlien's forty acres from his father's (Thure's) forty. The child in the photo, Charles North, would later become the uncle of Sterling North, who wrote *Rascal*. WHI IMAGE ID 147684

working for H. P. Leavens Company, a manufacturer of flour sacks and twine as well as a printing service, located on Water Street.

On January 15, 1883, Ludwig, who had connections in the Wisconsin Natural History Society, wrote to his father from Milwaukee: "Let me know at once if you would accept a position as custodian of the old [German-English Academy] Museum, now in the hands of the city."[5] Kumlien quickly responded that he would accept the position and was offered a one-year contract and $720 a year to become the Milwaukee museum's conservator in charge of collection management, though his main employment at the time was taxidermy. The position included time in the spring to remain at Koshkonong and collect for the museum.

In her history of the Milwaukee Public Museum, Nancy Oestreich Lurie wrote that Kumlien's "title as 'conservator' rather than curator reflected the use of continental European terminology in the museum's early years."[6] That the European-trained Kumlien, who had been collecting American birds for forty years, played an important role in the "continental European" organization of American specimens into a museum collection in Wisconsin was fitting. And it's a good story. There are two traditions of natural history collecting in this story, and Thure Kumlien embodied them both. And there was one man who had the perspective to understand both: William Morton Wheeler. He was there too, at the end, to tell of it.

As soon as Increase Allen Lapham arrived in Wisconsin in 1836, he began collecting and documenting the plants, minerals, and more that he found on Milwaukee's lakeshore and along its rivers. Quickly amassing a sizeable collection and always striving to use his knowledge to educate, Lapham began to make inquiries among men in the East as early as 1843 about establishing a museum of natural history in Milwaukee. With limited means, and with his efforts spread thin among many areas of interest—among them the education of women and the founding of the Milwaukee Female Seminary—nothing came of Lapham's museum plans, but his personal collection continued to grow. Lapham had an office and a cabinet of curiosities near his home on Chestnut Street in Milwaukee to house his growing collection.

Kumlien and Lapham were both "field men" in the sense that they spent a great deal of time hoofing it through the relatively unspoiled beauty of Wisconsin, observing and collecting. Unlike Kumlien, Lapham did not sell much of what he collected to scientists in the East; instead, he exchanged with them, especially plants, shells, and minerals. Lapham wrote a number of practical and informative pieces for publication. Though Kumlien wrote little, what he did write was more elegant than Lapham's prolific and utilitarian prose. As naturalists, both were singular and unique in their immediate worlds, but they corresponded with fellow naturalists in the east. If they had known each other, and there is no evidence that they did, they likely would have understood each other's curiosity and intensity and practicality. Both were scientists of their age and cared deeply about identifying Wisconsin's plants and animals. To Kumlien, especially, "the accurate systematic comprehension of the species was the alpha and omega of biology."[7] Both men were trying to answer the questions: What is here in Wisconsin, and what are the correct names for what is here? This New World practicality and desire to answer the most basic questions was one of the traditions at play in the story of the Milwaukee Public Museum.

The other, European, tradition in the museum's story was embodied by Peter Engelmann, who arrived from Germany in 1849. Educated at uni versities in Berlin and Heidelberg, Engelmann was one of the "Forty-eighters"—liberal German intellectuals who arrived in Milwaukee in 1848 and 1849—who brought to the city what Wheeler would later call "a super-heated atmosphere of German Kultur."[8] In fact, it was a superheated atmosphere very similar to that of Uppsala University in the 1840s. Both cultures were manifestations of the same liberal and Romantic movements sweeping Europe.

Engelmann, like Lapham, had a passion for natural history and for education. In 1851, he founded the German-English Academy in Milwaukee, including in his curriculum field trips for the study of plants and animals, geology, and archaeology. And he built at the school a cabinet of specimens, which first took up part of a classroom but then grew to fill the whole room. A number of men, among them Carl Doerflinger, volunteered to classify and preserve the specimens at the German-English Academy, usually called the German Academy. These men created the *Naturhistorischen Verein von Wisconsin*, the Natural History Society of Wisconsin, in 1857. Centered at the heart of German intellectual life in Milwaukee, this natural history society flourished.

Increase Lapham had organized the *first* Wisconsin Natural History Society in his Milwaukee office on March 3, 1848. Not enough members attended these meetings, and it petered out. Apparently, "the members, business and professional men, rode a single scientific hobby and would not attend a meeting when a paper was presented on a subject in which they lacked either knowledge or interest."[9] Lapham, interested in everything, could not understand this.

Language and cultural differences kept a distance between these two natural history societies of Wisconsin, yet there was friendship and cooperation among some members—Engelmann, Doerflinger, and Lapham.

All of Milwaukee and much of Wisconsin worked in 1865 to raise money through the Wisconsin Soldier's Aid Fair to build in Milwaukee a national soldiers' home for injured veterans of the Civil War. Lapham donated a valuable collection of natural history specimens to become the property of whichever Milwaukee educational institution was deemed most popular by raising the most money. The competition soon narrowed into a contest between Milwaukee's two favorite schools, Engelmann's German-English Academy and the Milwaukee Female College that Lapham had helped found. During their cooperation while raising money for the soldiers' home, Lapham and Engelmann became good friends and then "exchange professors" in each other's institutions.[10] In gratitude for Lapham's generosity to the German natural history collection, Doerflinger gave Lapham a good piece of the famous Trenton meteorite found on a farm in Washington County.[11] And in return, a part of Lapham's natural history collection became part of the natural history collection of the German Academy.

After Engelmann's death in 1874, the German Academy collections continued to grow. But after Lapham's death in 1875, his collections were sold to the University of Wisconsin in Madison.

The Milwaukee Public Museum was established when the collection of the German Natural History Society ran out of room in the German Academy. A group of volunteers, who had worked at classifying and preserving the academy's collection, organized and incorporated the museum in 1882, electing the first board of trustees and electing Doerflinger first "Custodian" of the museum at an annual salary of fifteen hundred dollars. The city institution was funded by an assessment of one-tenth of a mill of every dollar collected in property taxes, connecting the growth of the museum to the growth of the city.

Lapham had been working toward a natural history museum in Milwaukee since 1843, the same summer that Kumlien had arrived in Wisconsin. It had taken forty years.

The first museum employees, hired in 1883, were forty-year-old Doerflinger, who had arrived from Germany when he was six; twenty-nine-year-old Carl Thal, who had recently arrived from Germany; and sixty-five-year-old Thure Kumlien who had come from Sweden the year Doerflinger was born. Doerflinger, a former student of Peter Engelmann's and a volunteer for the Natural History Society, had urged for the creation of a museum he wanted to call the Lapham-Engelmann Institute. Serving in the Wisconsin Twenty-Sixth in the Civil War, Doerflinger had been seriously wounded at Chancellorsville, Virginia. His leg was amputated, leaving him with life-long pain. Thal was hired to be Doerflinger's assistant, as well as a janitor. But for a time, Thal did almost everything at the museum except science. He was secretary and office manager, and eventually he became the museum's librarian—all for five hundred dollars

a year. Kumlien's job title was "taxidermist and conservator."[12] His contract, as well as those of all museum employees, came up for renewal every
year.

Kumlien apparently did not hesitate to accept this position in Milwaukee, but he had to be coaxed to actually leave home for it. For forty years,
he had lived simply in the quiet woods where he knew all the birds and
plants. Though he had visited Milwaukee and seemed to enjoy it, Kumlien
was reluctant to leave the country and kept coming up with reasons to stay
home: he was ill, it was too hot to travel, his sister-in-law could not be left
alone, the snipes were just about to arrive. When the lake is high, he wrote
in a letter to Doerflinger, the chances are greatly in favor of the snipes and
against him. In the fall of 1883, Kumlien told Doerflinger that he would
look for a place near the museum where he could board, a place for a man
"used to frugal living and natural habits."[13]

In Milwaukee in late 1883, preparations were being made to move the
natural history museum from the German Academy to the Milwaukee
Industrial Exposition Building. The city had leased eight thousand square
feet in the newly completed and mostly empty building for eight hundred
dollars a year. The building was to be divided into work space and two large
exhibit halls named the Lapham Room and the Engelmann Room. The
Lapham Room already contained an exhibit by Henry A. Ward's Natural
Science Establishment—a commercial exhibitor that also supplied natural
history specimens to schools and colleges. But the Ward exhibit had to be
moved so the renovations could be completed on the two exhibit rooms.
While Doerflinger and Thal were overwhelmed with hundreds of tasks,
Kumlien was still at Koshkonong.

F. C. Winkler of the museum board wrote Kumlien on December 15,
explaining the procedure of moving the cases from the academy to the
Engelmann Room, which entailed disassembling the cases for safe trans-
port.[14] New glass cases had also been ordered and would arrive in a few days,
so the contractors were working hard to prepare the rooms. When some
of the new cases arrived, the museum staff would have to put the stuffed
animals from the Ward collection into these cases to keep them away from
construction dust. Before they moved everything to the finished room,
Winkler wanted Kumlien, with the assistance of a young man Kumlien

This H. H. Bennett photograph shows the Milwaukee Industrial Exposition Building a few years after it was completed in 1880. The Milwaukee Public Museum rented space here until 1898. The building burned in 1905. WHI IMAGE ID 7440

would bring with him from Koshkonong, to examine and brush and dress the plumage and pelage of the animals. No dirt or rubbish or bug-infested specimens should be put into the museum's splendid new glass cases.

Kumlien, still in Busseyville, wrote to Doerflinger on December 18, explaining that he was busy packing the birds Doerflinger had asked him to send.[15] On December 20, Doerflinger fired off a telegram to Kumlien: "Letter received. Terms satisfactory. Come soon as possible."[16] By January, Kumlien was in Milwaukee working with Doerflinger and Thal.

Kumlien was essential to the moving of the bird, mammal, and plant specimens from the academy collection to the Milwaukee Public Museum. He was the only one in the city and the state with the ability to identify all the unlabeled or mislabeled bird and mammal specimens, mount the new specimens, and repair the damaged ones. And he had an eye for the aesthetics of the display. With the help of two part-time taxidermists, Kumlien mounted forty-nine bird skins for the new museum.

Kumlien also had forty years of experience in finding and treating insect infestations in specimens. Every bird and mammal in the collection had to be treated with corrosive sublimate. In his 1883 report of work done, Doerflinger wrote, "The bisulphuret of carbon and other drugs and poisons used, certainly made the museum a very unhealthy abode for human beings."[17]

In February 1884, Kumlien was again hired for a year by the museum's board of trustees as taxidermist and conservator at a remuneration not to exceed fifty dollars a month for an indefinite period.[18]

William Morton Wheeler had been on hand as the new museum was being established in January 1883, and he donated his large collection of beetles to the museum. Not yet twenty years old, Wheeler had been a student of Peter Engelmann's at the German Academy and had "haunted the old academy museum since childhood and knew every specimen in it."[19] Wheeler wrote, "Of course, I was on hand when Professor Ward's boxes arrived, and I still remember the delightful thrill with which I gazed on the entrancing specimens that seemed to have come from some other planet. I at once volunteered to spend my nights in helping Professor Ward unpack and install the specimens, and I worked as only an enthusiastic youth can work."[20] Impressed by this "highly gifted son of science,"[21] Ward offered Wheeler a job in Rochester, New York, at his Natural Science Establishment, which Wheeler described as "not so much a museum as a museum factory."[22]

In March 1884, four plumbing leaks in an upper-floor bathroom caused the loss of four small mounted mammals in Lapham Hall. Three hundred dollars' worth of stuffed specimens were saved by Kumlien's prompt and skillful treatment.[23]

When the Milwaukee Public Museum opened to the public on schedule, May 24, 1884, more than a thousand people came that first evening to hear the speeches, to peer up at Ward's dinosaur skeleton, and to shuffle past exhibits of minerals, fossils, plants, mammals, the birds of Wisconsin, reptiles, fish, bugs, and coins.

By the summer of 1884, Doerflinger had come to rely even more on Kumlien. When he had to be away for a time in June, Doerflinger wrote

This room in the Milwaukee Public Museum held exhibits by Henry A. Ward's Natural Science Establishment at the time Kumlien worked there. Part of his job involved sitting in the exhibit halls and answering museum goers' questions. WHI IMAGE ID 7498

Kumlien that he was glad Kumlien would be at the museum: "I shall feel at ease in regard to the necessary 'mothering' or 'fathering' of our menagery."[24]

Several terrible fires of this era must have made the museum staff in that large wood-frame building skittish. They certainly made sure their fire insurance was in order. The Newhall House Hotel, on the corner of what is today Broadway and East Michigan Street in Milwaukee, had burned on January 10, 1883, killing more than seventy people. Then the Science Hall fire at the University of Wisconsin in Madison destroyed Increase Lapham's collection of fossils, shells, and more on December 1,

1884. Lapham's pressed plant specimens were in another building and still exist in the University of Wisconsin's herbarium.

In 1883 and 1884, Wheeler was in Rochester working at Ward's Natural Science Establishment, identifying birds and mammals, listing collections of shells and sponges, and preparing catalogs and price lists for publication.[25] While Wheeler was at Ward's, a farm boy named Carl Akeley came to study taxidermy with Ward's taxidermist in 1883 or 1884 and quickly learned all the elder taxidermist had to offer. Akeley, according to Wheeler, was "born with unusual taste and discrimination and an intuition which could dispense with mere book-learning. Of all the men I have known, . . . he seems to me to have had the greatest range of innate ability."[26] A close friendship developed between Akeley and Wheeler—they were both the sort who, Wheeler wrote, "soon exhaust the possibilities of their medium, like fungi that burn out their substratum."[27] The two friends decided to leave Ward's, which Wheeler felt was "after all, neither an art school nor a scientific laboratory, but a business venture."[28] At the end of June in 1885, Wheeler returned to Milwaukee to teach German and physiology at Milwaukee High School. Akeley stayed on at Ward's for a few more months, taking advantage of the rare opportunity to assist in skinning and mounting an elephant—Jumbo, P. T. Barnum's famous circus elephant, which had been hit by a train. Eventually, Akeley joined Wheeler in Milwaukee, where Akeley set up his own taxidermist shop.

Wheeler immediately began collecting plants and insects and writing up everything he saw. After being elected to the Natural History Society of Wisconsin in November 1885, he quickly presented papers on the flora of Milwaukee, the distribution of beetles along the Lake Michigan beach, and the trees in the city of Milwaukee. Fluent in German and other European languages, as well as in the language of natural science, the observant, enthusiastic, and energetic Wheeler made himself useful in the Natural History Society. Later, when he had done all he could in Milwaukee and at the museum, Wheeler went on to the University of Chicago as a student and then professor, followed by professorships at the University of Texas and Harvard. Wheeler became the most important entomologist of his time.

Even before Wheeler came to work officially for the museum, Kumlien would have known him. Wheeler, with an education from the German

This photo of William Morton Wheeler was taken around the time he became the second director of the Milwaukee Public Museum in 1888, when he was twenty-three years old.
MILWAUKEE PUBLIC MUSEUM

Academy similar to the one Kumlien had received at Skara Gymnasium, might have reminded Kumlien of himself at that age. Wheeler quickly came to admire Kumlien, but to Wheeler, who was born in 1865, Kumlien was a naturalist of the old school. "In my many talks with him," Wheeler wrote of Kumlien, "I well remember his looks of wonder whenever I touched on some of the embryological and morphological problems of the day. The advance of biology along a path so different from the one he had followed since he listened to lectures in Upsala, never failed to astonish him."[29]

Stimulated by the company of Wheeler and other German-educated men, Kumlien brushed up on his German. He wrote his daughter, Swea, "I have a good opportunity to improve my old zoological brains by reading books in that line and would you believe it, I have read through three volumes of a German manual for preserving all sorts of animals. At first each page took some time but it goes faster now."[30] But Wheeler noticed Kumlien look askance at young naturalists who misused the Latin he so loved. Wheeler wrote that Kumlien "fondly loved the simple Linnean names, which cling fast to the memory, and often wondered how some modern systematists could fall into awkward and inappropriate nomenclature."[31]

In his *The Flora of Milwaukee County*, Wheeler credits Kumlien with two finds. Right next to the Exposition Building where he worked, Kumlien had found *Euphorbia peplus*, a European spurge. And when he wandered south of downtown Milwaukee, along the railroad tracks to Bay View, perhaps on his way to see Ludwig where he lived on Scott Street,

This photograph capturing Milwaukee's German Market Hall (the structure in the distance on the right side of the street) was taken by H. H. Bennett in the early 1880s. As the market was not far from the Milwaukee Public Museum, Thure Kumlien might have wandered this area looking for some familiar northern European food.
WHI IMAGE ID 7487

Kumlien had found a specimen of coast knot grass.[32] Though we can't prove it, Wheeler and Kumlien must have wandered the city together, Wheeler showing Kumlien places he had known as a boy. Wheeler certainly knew that Kumlien was "a singularly acute observer, though his powers of observation were stimulated by an intense and childlike love of natural objects rather than by any interest in their importance from a speculative standpoint."[33] Wheeler's interest was in the speculative, while Kumlien's interest was in the beauty of the object itself. As a result, Wheeler continued, Kumlien's conversation "teemed with interesting facts, but seldom rose to wide generalizations." Kumlien was "satisfied to observe, to collect and prepare plants and birds because they were full of marvelous beauty and offered endless material for comparative study."[34] Wheeler, who wrote these remarkable impressions of Kumlien when he was in his twenties, was an astute observer of human nature and also must have spent a considerable amount of time with Thure Kumlien.

While he probably walked the city streets with Wheeler, Kumlien most certainly explored Milwaukee on his own. He may have spent some time looking for the Milwaukee of 1843 that he had seen with Christine. He may

Though the Greek Revival courthouse where he and Christine
were married in 1843 had been torn down in 1870, Thure may have
appreciated the grander Roman Revival courthouse pictured here.
You can faintly see it through the trees in this H. H. Bennett photo
from the 1880s. Thure, a Romantic flute-player, would have appreciated the formality of this fountain with its flute-playing water
nymphs and winged cherubs. WHI IMAGE ID 148054

have gone down to the harbor to look for the Huron Street Pier where they
had first set foot in Wisconsin together, but that pier was gone and the
harbor must have seemed an unrecognizable bustling place to Kumlien.
He may have looked for the attractive white frame courthouse where he
and Christine were married. But an ornate stone courthouse had since
replaced it. In front of the courthouse now was a garden with a fountain
and benches where Kumlien could sit to collect his thoughts and listen to
the water. He may have gone looking for the ducks and geese in the wild
rice marshes and the tamarack swamps, which he had seen in Milwaukee
in 1843, though even then they were disappearing. By the 1880s, they were

filled in, gone. Industries on the Rock River Canal and the Milwaukee River were fouling the waters with the refuse produced by tanning hides, brewing beer, and turning trees into board feet. The lake bluff had been scraped of its old forest trees. Yet Kumlien would likely have admired the new ornate buildings—the Mitchell Building and the Grain Exchange at Water and Michigan Streets, the new seventy-bed Lutheran Hospital called Passavant. With his habit of finding whatever beauty there was, Kumlien must have seen plenty walking on the wide, dusty streets of the growing city.

One evening during his time in Milwaukee, Kumlien wrote a letter to Swea telling her about his typical day. He said he left work at the museum at five thirty and went to get his supper, which was often just coffee, "after which till bedtime I have nothing to do and nowhere to stay unless I go up to Ludwig's—2 miles from here and that I do sometimes, but I can't go there every evening and a saloonist I am not, so I have to go up to my room and what to do there? Go to bed on those wooden slats—no not until nearly nine o'clock," though he said, "my hind quarters are getting more used to the torturing invention" of the thin mattress on slats. "But," he continued, "I can write and I have many to write to. Tonight I find my thoughts stretching out very strongly toward Cedar Mills, Minn.," where Swea lived with her husband and their children, Edna and Winfred, born in 1883 and 1885.

"I am well," he went on, "and it is with real pleasure that I am able to tell you so—I am old you know and am fast approaching the three score and ten which is considered a ripe old age."[35] Kumlien also wrote to Theodore at the farm asking him to trace the feet of little Madge, Charlie, and Angie and to send the "shoe measures" back to him in Milwaukee so he could have shoes made for his grandchildren. Sophia Wallberg sometimes lived in the same Milwaukee quarters as Kumlien when she occasionally worked as a cleaner at the museum.

On April 30, 1885, when Kumlien was at home in Koshkonong for the spring collecting, he wrote his friend Doerflinger with concern about his health: "You must be careful in this weather." Kumlien reported that he had arrived home on Wednesday evening at about eight o'clock, "footing it from Edgerton." The next day it rained, but he said, "I cleaned up the gun & cut wads, etc. Friday I . . . tried the gun and 3 good skins was the fruit of the 3 first shots, not that I don't miss sometimes, but I find that I have not all together forgotten how to handle a gun."[36] Kumlien found that with no one living in his house for four months, the mice had girdled his young apple trees, giving him a new appreciation for cats.

That April evening, Kumlien wrote to Doerflinger one of his most beautiful descriptions:

> The whippoorwill however has come. . . . I don't know what makes me like that bird so much if it is not because it was a favorite with my wife. Of course "sua cuique mos est" ["to each his own"] with birds as well as folks, but we used to think that it was something peculiarly lovely in the confiding manner in which the whippoorwill, noiseless as a dead leaf, used to drop down on our piazza or by the door, within a few feet of us and give us all the music she had and that at a time of the day when the rest of nature was silent, excepting perhaps some frogs croaking down below in the marsh.[37]

Back in Milwaukee for the summer of 1885, Kumlien worked. With the occasional help of another taxidermist, Kumlien stuffed and mounted forty-five animals, mostly birds. He prepared 258 skins of birds and other animals to be kept or traded as duplicates. He remounted 750 stuffed specimens on new ash stands, or "tablets," made there in the workshop so they were consistent. The standards—the branches on which the birds or mammals were perched—were gathered from trees near the city. Before they were used, to rid them of insects, the branches had to be soaked in an arsenic solution. Kumlien cleaned and repaired the mounted birds or mammals if necessary. He also inspected the condition of the entomological collection. The labels of nearly all the birds and mammals were reexamined in the light of the latest information available. That summer, Kumlien frequently noticed the inadequacy or inaccuracy of records of specimens that had been received at the museum in times long past. Somehow, by the end of 1885, Kumlien had found time to collect 180 birds and mammals and a large number of fossils and mineral specimens from Koshkonong[38]— all of which he donated to the museum.

Besides all this, during visiting hours at the museum, Kumlien was often available to greet visitors and answer questions. On October 4, Kumlien wrote to Theodore: "Yesterday afternoon 2300 visitors in the museum and in the evening 1200, the largest attendance since the museum was opened. Yesterday over 4,000 school children were given a free chance to see the exhibition!"[39]

For more than thirty years, since Kumlien and his young neighbor Edward Lee Greene had botanized in the marshes and woods, young people had been attracted to this man who some described as "childlike." They trusted him and recognized in him their own uninhibited curiosity. As his friend Wheeler said, the fact that Kumlien genuinely enjoyed the company of young people and freely gave them the benefits of his knowledge and experience "is not to be underestimated in a naturalist, for they are the means of charming the young and making good naturalists of youths" who would be repelled by cold or heavy-handed authority. Wheeler added, "Many of our rising botanists and zoologists owe much to Mr. Kumlien's warm and sympathetic enthusiasm."[40]

In the fall of 1885, the acting president of the museum reported that "the collection of birds has improved materially in appearance and value under the loving hands of our able ornithologist Mr. Thure Kumlien, and the Wisconsin group has made an appreciable advance toward completion."[41] Kumlien had an exceptional range of abilities that the museum valued. He was skilled with his hands, artistic, investigative, and social. Farming had forced Kumlien into a conventional role that had not suited him. He was not the enterprising sort and had no desire to sell or promote himself. In the 1880s, he seemed to enjoy the stimulation of the city and the museum and the people he met as well as the quiet and peace of his cabin in the woods. He had the best of two worlds.

Just a few years after its opening, the museum was successful. Attendance was high and donations of objects keep coming in—a hummingbird nest on a branch donated by Julia A. Lapham (the daughter of Increase Lapham), donations from Charles Mann, and butterflies from Dr. Philo Hoy. Wheeler donated collections of snails and insects from Rochester, New York.[42] Also donated was a collection of bird skins from other parts of the world, including from South America.

It must have been a heady time at the museum when Wheeler, Doerflinger, Thal, and Kumlien were working together. When Akeley arrived in 1885, Kumlien described, with admiration and no envy, Akeley to Theodore: "Young Carl Akeley, about whom I think I have spoken to you, is to do the principal taxidermy, and he is, without question, the ablest taxidermist I have ever seen."[43] Beginning with the muskrat family at the Milwaukee Public Museum, Akeley would eventually revolutionize the presentation of mounted animals in naturalistic habitats, first at the Milwaukee Public Museum, then at the Field Museum in Chicago and at the American Museum of Natural History.

It must have been after the arrival of Akeley at the museum that Kumlien wrote the following to Swea:

> You wonder what I do in the museum all day—stuff birds? No not very often—I have only stuffed about a dozen and most of them East India beauties since I came here this last time. No—today for instance I picked out the best specimens of fossils from a lot which D[oerflinger] brought from 300 miles north of here about three weeks ago. In the afternoon I don't do anything but answer many questions in regard to specimens in the museum and that isn't very hard work.[44]

In December 1885, Edward Lee Greene wrote to Kumlien from Berkeley, California. Greene had continued writing to Kumlien since his first botanical letters as a teenager in the Civil War. Now Greene was a prominent botanist at the University of California, but he still wanted to tell his old teacher of his successes. A year earlier, he had sent Kumlien several monographs about more than fifty new species, but Kumlien had not responded. In the December 1885 letter, Greene crowed a bit about how he and botanist Asa Gray at Harvard had been coincidentally working on the same plant, but Greene had named it in print before Gray. "The best thing of all," Greene wrote, was that he had found a *Ranunculacea*, a type of buttercup, in the mountains of California and had named it *Kumlienia*—"if it please you!"[45]

Around this time, Kumlien wrote to Greene a letter "tinged with melancholy," informing him that the trees in their cherished tamarack swamp had been cut down. The blueberries and cranberries were gone, and so were the orchids. Kumlien told Greene that men "of the common sort, had drained it and ploughed it and planted it with market-garden vegetables."[46]

In 1886, Ludwig Kumlien temporarily joined the staff of the Milwaukee Public Museum as a "determining collector." In 1887, Thure donated three

Visible in this 1883 photo of the frame house Thure Kumlien built in 1874 is the family's log cabin, which had been moved from its original site in the woods, at the right of the photo. On the porch are Sophia Wallberg and Theodore Kumlien. Frithiof Kumlien wrote to his father from this porch in the summer of 1887. WHI IMAGE ID 147683

hundred species of plants to the Milwaukee Public Museum, plants that were mostly new to the collection, and Ludwig donated ten arctic plants that he had collected on Cumberland Island off Greenland in 1878.[47] The energy of the three young men—Akeley, Greene, and Ludwig—may have reminded Thure of his advancing age.

Frithiof was at home on the farm in July 1887, between sessions at the Layton School of Art, when he wrote to his father in Milwaukee. "Sunday afternoon on the old porch. Beautiful beyond all description." He was working on a heron painting, playing guitar, and visiting neighbors, and he was very pleased with his progress on a painting he called "Potato Harvesters" in which he "never got such a good expression or effect before." He wrote his father that he wished "you were here to criticize but I must be my own critic." Frithiof hoped Ludwig could remain in the museum, as that would be such a fitting position for him, and he sent love from Moster. "The folks say I am better than they ever saw me," he concluded, "Good bye Frithiof."[48]

Kumlien quickly wrote back to his youngest son: "As I told you before there are hosts of beautiful things in the world, that are intended to elevate our minds and add to our pleasure if we only see them and I am glad to see that you not only have your eyes open for them but that you seem to enjoy some of them at least."[49] Many fathers would have worried about a son pursuing a career as an artist rather than in the field of pharmacy, in which Frithiof had a degree. But not Thure Kumlien. He wrote, encouragingly,

> Now I am glad that you seem to be in your real element and I repeat that if you haven't the materials suitable please let me know—for now you must be helped and no mistake, as yet you cannot be expected to have arrived at the place where your productions can compete in the market with so many others, just dig and dig and dig and you will find just where your strength lies and we will by and by see that you can get suitable instruction—in the meantime work—work and draw and paint and draw as I have said before—anything, so that you get so that you can readily copy anything! The *painting part* depends so much *on correct drawing*."[50]

In the fall of 1887, over a period of months, Thure and Ludwig worked together to accomplish "an important piece of work of permanent value" to the museum. They made the ornithological collection "a thoroughly reliable source of information, and a credit to the institution." Ludwig put together in the form of a catalogue all the birds in the collection with descriptions and remarks. In his annual report to the board of trustees, Doerflinger wrote, "The collection is now rearranged in ten glass cases, according to the new system of classification, beginning with the wingless birds, and the contents of each case indicated by conspicuous card-board signs." When Thure and Ludwig weren't sure of the identity of a bird, they sent either the bird skin or a drawing of the bird to Professor Robert Ridgway at the Smithsonian or Ludwig's friend Joel Asaph Allen, the first curator of birds and mammals at the American Museum of Natural History in New York, asking for help in identifying them. Then the father and son team began to do the same with the collection of mammals and reptiles. Doerflinger wanted to hire comparable specialists to do this for the remaining departments of the collection, so the institution would "Obtain its highest usefulness for the purposes of education and science."[51]

That fall, the more than twenty thousand zoological specimens in the museum were valued at almost twenty thousand dollars, nearly half the aggregate value of the contents of the museum. The nearly seven thousand botanical specimens, many of which Thure and Ludwig had either worked on or contributed, were valued at almost a thousand dollars.[52] Even in just monetary terms, the Kumliens—mostly Thure—had made enormous contributions to the early Milwaukee Public Museum through their donations and their "re-examination, relabeling and mounting of many thousand specimens."[53]

This photo of Ludwig Kumlien, ornithologist and oldest son of Thure Kumlien, was taken in 1891. WHI IMAGE ID 147779

Doerflinger resigned his position as custodian on October 1, 1887, because of ill health. Twenty-two-year-old William Morton Wheeler was appointed custodian in his place.

On January 6, 1888, in Milwaukee, Frithiof Kumlien died unexpectedly at the age of twenty-eight. He died in his bed, likely of heart disease. Frithiof was buried at Sweet Cemetery next to his older sister Agusta and his mother, Christine.[54]

His father and brother apparently went home to bury Frithiof and then came back to Milwaukee to work. But shortly after that, Thure collapsed. He was likely sickened by grief and, perhaps, by the effects of prolonged exposure to arsenic. Ludwig tried to take his father home to Koshkonong, but Thure was so ill that it took Ludwig two days of intermittent travel to get him home from Milwaukee. From Koshkonong, Ludwig wrote to Doerflinger: "He has been gaining steadily but slowly since he came home. . . . His appetite has returned. He is yet very weak and I am quite unable to say just when he will return to Milwaukee. I can leave him safely next week and will then return to Milwaukee. He has been a very sick man."[55]

Thure recovered enough to go back to Milwaukee to work. He was working at the museum in May when Theodore and Alice's fifth child, a son, was born at the farm. At first, they named the baby boy Wendell Frithiof Kumlien, which Theodore wrote in a letter to his father. Madge, who was eight, wrote her grandfather a letter, as well.

On July 20, Thure wrote Theodore at the farm from Milwaukee: "My dear Theodore: Thanks for your letter! I need not tell you how good it feels to read 'We are all well.' My pleasure at receiving a genuine letter from Madge you can easily imagine! I have just now not time to write to her, but I will soon do it. Please tell her that it did me a heap of good to receive this first real letter from a grandchild of mine." He announced that he had been hired again to work at the museum until at least the first of May—"to do some taxidermy and to collect, press, and dry plants."[56]

Thure continued about the new baby: "Now about the name of the *boy*! Well what to say? That you called him Frithiof pleased me more than I can say! I do with all my heart thank you for that! . . . But what harm can it do to put an L. either before or after the Frithiof giving him three names [in the Swedish way]?" Thure's given name was Thure Ludwig Theodore Kumlien. He continued:

> At the same time Frithiof must not be omitted under any consideration, since you gave him that name, which for me, who knew him, is so very, very, dear. I consider the little fellow a providential make-up for what we lost and I cannot help but think a good deal of him, little as he is, and it does me good to hear that he is doing well. . . . Love to Alice and all the children, from your old and affectionate Father. PS Tell Moster that I am in pretty good health; I am in better flesh than I have been for a long while.[57]

Theodore and Alice changed the baby's name to Wendell Frithiof *Ludwig* Kumlien.

Though Thure claimed to be in good health in his letter, Ludwig later wrote to Greene that since Frithiof's death, his father had hardly been himself, as he took it very hard. And Swea was very sick, as was her husband. This, Ludwig said, "worked upon his mind a good deal."[58]

According to Ludwig, his father had saved enough money to live comfortably on the interest and intended to resign in September.[59] Ludwig also reported that near the end of July, Thure had inhaled "too much corrosive sublimate" when he "poison[ed] a lot of plants."[60]

Otto Ernst took this photograph of Thure Kumlien at his work table at the Milwaukee Public Museum in 1887. WHI IMAGE ID 69876

On Sunday, August 5, 1888, Thure Kumlien brought up from the basement of the museum a large collection of South American bird skins that had been stored for a time. While examining the birds, he inhaled the poison preservative and became violently ill. He was immediately taken to the nearby Passavant Hospital and cared for by a skilled gastrointestinal physician named Dr. Nicholas Senn.

But Thure Kumlien died at three thirty that Sunday afternoon.

Wheeler wrote the museum's insightful and moving memorial to Kumlien. Though he was only twenty-three, Wheeler was up to this task. The Sixth Annual Trustees Report concludes with Wheeler's black-bordered, ten-page obituary of Thure Kumlien, which has been quoted from extensively in this chapter. Wheeler, who was brought up in the same scientific and Romantic tradition as Kumlien, wrote: "Mr. Kumlien was no narrow man. He was passionately fond of painting, music and poetry. I have heard him repeat with a glow of delight verses from Runeberg" and from Tegnér's *Frithiof's Saga*, "rendering the wonderful rhythm of the latter with exquisite grace and precision. He was a man of most refined tastes, without any of the extravagant desires which such tastes often engender. He was satisfied to live most simply a life which most philosophers might envy."[61]

Angie Kumlien Main remembered when her grandfather came home to Koshkonong for the last time: "He was in a big black buggy with fringe hanging from the top and plumes on the corners. We could look in through the glass which enclosed his 'buggy' and see a long dark box."[62] Thure Kumlien was buried at Sweet Cemetery alongside Frithiof, Christine, and Agusta.

Until she was fourteen and the family moved to Fort Atkinson, Angie Kumlien visited her great-aunt Sophia Wallberg in the old cabin in the woods several times a week. She would walk back through their farmyard and follow the little road along the long row of Indian mounds. For most of the way, she could see Lake Koshkonong, except where a small piece of woods the family called "the breaking" hid the view. The breaking was the piece of woodland Thure Kumlien had saved by fencing it off, never letting cows or pigs graze there. He had left it as he had found it in 1843, lush with maidenhair ferns and yellow lady's slippers.

When Angie would get to the cabin, Aunt Sophia would let her and the other children play with the tall mounted birds, as long as they put them back again in the shed. The children played horse with the pelican, the great blue heron, and the sandhill crane.

By the 1880s, Thure Kumlien had seen market hunting, the drainage of wetlands, and other loss of habitat force the sandhill crane, left, to near extinction. Since passage of the Migratory Bird Act of 1916, the Migratory Bird Act Treaty of 1918, and other protections, the sandhill crane is becoming again a familiar sight in Wisconsin. This sandhill crane was painted by John James Audubon in the winter of 1832–1833 in Boston using a live crane as a model. Audubon's background, which looks like white mountains, is intended to be Florida sandhills. He identifies the bird as a "Hooping Crane." In the early days of American ornithology, both Audubon and Alexander Wilson believed sandhill cranes to be young whooping cranes. NATIONAL AUDUBON SOCIETY

Epilogue

In the month after Thure Kumlien died, Ludwig wrote to his friend and botanist Edward Lee Greene in California. Greene was gathering information from Ludwig to use in his memoir of his teacher for his new botany journal *Pittonia*. Ludwig wrote, "Father and I were jointly writing a book on the birds of Wis. embodying his observations for 40 odd years and I must finish it."[1] Ludwig and Thure's exhibit of the birds of Wisconsin for the Milwaukee Public Museum had been a draft of that book.

Ludwig did finish the book, with ornithologist Ned Hollister, when Ludwig was a professor at Milton College. *The Birds of Wisconsin*, which included many of the forty years of Thure Kumlien's observations, was published first as a bulletin of the Wisconsin Natural History Society in 1903. Sadly, Ludwig died of cancer in 1902, before its publication. Ludwig left behind his widow, Annabel Carr Kumlien, and two young children. (Gregg Kumlien, whose painting appears on the dedication page, is the great grandson of Ludwig.)

Theodore's second daughter, Angie Kumlien Main, preserved her grandfather's papers and wrote a three-part biography of him, along with other articles, keeping his name alive. Because she saved and donated his letters and bird lists to the Wisconsin Historical Society Archives, we can read most of the letters from Brewer, many drafts of Thure's letters to Brewer, and other letters besides. Angie also saved and donated lists of the birds Thure sent to museums in this country and in Europe. And she wrote her own "Birds of Wisconsin." Her *Bird Companions* combined poetry with her own observations of birds, and it included frequent mention of Thure's observations on the birds of Koshkonong.[2] Many of his mounted birds and a few of his mammals were given to the Hoard Museum in Fort Atkinson. Still today, we can visit the museum's bird room on the second floor and peer into cases of birds mounted by Thure Kumlien.

The thousands of study skins Kumlien provided to museums still exist, in large part. As scientists develop new ways to learn from study skins in museums, such as extracting DNA from them, these specimens will have value Kumlien never imagined. Because he walked the original landscape

Though rare, *Cypripedium reginae* is still found in tamarack bog remnants in Jefferson County. This showy lady's slipper was photographed by naturalist Aaron Carlson, one of the volunteer rare plant monitors for the State of Wisconsin Department of Natural Resources. PHOTO BY AARON CARLSON

known and enjoyed by American Indian peoples for generations, Kumlien's specimens provide a special window onto Wisconsin's past. His collections have provided us with important baselines to be used not only in ornithology and botany but also in conservancy and restoration. His letters, lists, specimens, and writings allow us to walk Wisconsin's vanished landscapes in our minds.

In this age of climate change and extinctions, perhaps the most important legacy of Thure Kumlien's life is that he has shown us the magnificence of the world we are trying to save. We can be strengthened in our struggles to save this world by holding in our minds what Thure Kumlien saw. He offers us "hosts of beautiful things in the world . . . if we only see them."

NOTES

Prologue

1. Adapted from Edward Lee Greene, "Sketch of the Life of Thure Kumlien, A.M.," in *Pittonia: A Series of Papers Relating to Botany and Botanists, 1887–1889*, vol. 1 (Berkeley, CA: n.p.), 250–60.

Introduction

1. Angie Kumlien Main, "Thure Kumlien, Koshkonong Naturalist," *Wisconsin Magazine of History* 27, no.1 (September 1943): 20.
2. Main, "Koshkonong Naturalist," 20.
3. Lorine Niedecker, *Collected Works*, ed. Jenny Penberthy (Berkeley: University of California Press, 2002), 145.
4. Walter Havighurst, *The Winds of Spring* (New York: Macmillan, 1940), 66–67.
5. Sterling North, *Rascal* (New York: Puffin Books, 1963), 55.
6. Rasmus B. Anderson and Albert O. Barton, *Life Story of Rasmus B. Anderson* (1915; repr. Charleston, SC: Hardpress, 2010), 92.

Chapter 1

1. Between Kolthoff and Kumlien were many similarities, though they grew up a generation apart. Both grew up within walking distance of Lake Hornborga in upper-class families. Both had passionate interests in collecting plant and bird specimens. Both had artistic abilities and romantic sensibilities. Both practiced taxidermy and later worked in natural history museums, with Kolthoff making major strides in the design of dioramas.
2. From Gustaf Kolthoff, "Mitt första besök vid Hornborgasjön 1861" ["My first visit at Lake Hornborga 1861"], in *Minnen från mina vandringar i Naturen* [*Memories from my walks in nature*] (Stockholm: Fr. Skoglund, 1897).
3. Kolthoff, "Mitt första besök."
4. Kolthoff, "Mitt första besök."
5. J-G Hemming, email to author, February 15, 2017.
6. Edvardsson, Aldste sonen I en barnaskara pa 14, Varldsberomd naturalist, Harlunda-Bjarka Hembygodsfornening, Arsskrift 2009, Argang 25.

7. Angie Kumlien Main, "Thure Kumlien, Koshkonong Naturalist," *Wisconsin Magazine of History* 27, no. 1 (September 1943): 21.

8. *Slakten Kumlien*, Svensk slaktkalender (1974): 215–19, from J-G Hemming, email to author, March 4, 2017.

9. A. B. Carlsson, *Uppsala Universitets Matrikel, 1750–1800* (Uppsala: Almqvist & Wiksells Boktryckeri AB, 1946), 43.

10. Petronella Johanna Rhodin Kumlien, "In Praise of the Maiden Life," trans. Thomas Edvardsson and J-G Hemming; Susanna Elizabeth Bjornsdottor Kumlien; and Veronica Lundberg; consolidated and edited by Martha Bergland.

11. J-G Hemming, email to author, March 3, 2017.

12. Quoted in Main, "Koshkonong Naturalist," 21. The source of her quotation is lost.

13. Main, "Koshkonong Naturalist," 23. No sketches by the parents have been found.

14. Thure Kumlien Papers, Wis Mss MQ, box 2, Wisconsin Historical Society Archives, Madison (hereafter cited as TK papers). Notes on the *Notbok* by MaryLee Knowlton and Holly Hetzter, email to author, January 19, 2017.

15. The Skara Cathedral School was called Skara Gymnasium by his granddaughter and biographer, Angie Kumlien Main.

16. The former high school building now houses an elementary school called Djäkneskolan, or Deacon's School.

17. Information on Skara from Gustaf Holmstedt, *Lekt och Lärt i Skara—Vad diarium, dagspress och dagböcker berättar om livet i Skara Skola, stad och bygd åren 1821–1850*, trans. and ed. Lena Peterson Engseth (Uppsala: Föreningen för svensk undervisningshistoria, 1983).

18. Holmstedt, *Lekt och Lärt i Skara.*

19. L. A. Cederbom and H. Vingqvist, *Lärare och rektorer vid Skara högre allmänna läroverk 1641–1941* [*Teachers and headmasters at Skara 1641–1941*] (Stockholm: G. Holmstedt, 1941).

20. Cederbom and Vingqvist, *Lärare och rektorer.*

21. Hanna Hodacs, "'Little Brother Carl'—The Making of a Linnaean Naturalist in Late Eighteenth Century Sweden" (unpublished manuscript, 2015, p. 4), quoted with permission of the author.

22. Jenny Beckman, "Collecting Standards: Teaching Botanical Skills in Sweden, 1850–1950," *Science in Context* 24, no. 2 (2011): 242.

23. Beckman, "Collecting Standards," 243.

24. *In the Field: Botany in the Wild* [Web exhibit], Botany Libraries, Harvard University, 2002 (last modified 2011), https://hwpi.harvard.edu/botlib_projects/book/field-botany-wild.

25. Wilhelm Luth, *Catalog Ofver Skara Kongl. Gymnasii Bibliothek*. (Skara: C.M. Tobin, 1830), 278; and Jean Anker, *Bird Books and Bird Art* (Copenhagen: Levin & Munksgaard, 1938), 171–72.

26. Luth, *Catalog*, 629.

27. Elis Erlandsson, *Skara högre allmänna läroverks lärjungar åren 1826–1869*, digital version, Project Runeberg; *Svenskt Porträttgalleri* [Swedish Portrait Gallery], digital version, Project Runeberg; Church Records, Steneby Parish, Älvsborgs län, *Fregatten Eugenies resa omkring Jorden åren 1851–1853*, 2 volumes, published by the Royal Academy of Science, 1855 and 1857, digital version, Project Runeberg.

28. Main, "Koshkonong Naturalist," 23.

29. Elsewhere in the letter, Petronella calls her sons "patients" and refers to a "dark situation." The boys may have been quarantined during an epidemic in the rooms where they boarded. Or it may be that this is a bad translation. The original is lost.

30. The grandmothers would have been Elin Landstrom, sixty-six, and Ludwig's mother, Maria Margaretta Hallarstrom Kumlien, seventy-seven.

31. Main, "Koshkonong Naturalist," 22.

32. Main, "Koshkonong Naturalist," 22.

33. Main, "Koshkonong Naturalist," 22.

Chapter 2

1. Women did not attend Uppsala University until the early 1870s.

2. Uppsala University, "A Historic Summary: The Period of Uppsala Romanticism, 1800–1877," https://uu.se/en/about-uu/history/summary/.

3. Bertil Nolin in Lars G. Warme, ed., *A History of Swedish Literature* (Lincoln: University of Nebraska Press, 1996), 151.

4. This was temporary; in a few years he had to borrow money to be able to leave Sweden.

5. Rasmus B. Anderson and Albert O. Barton, *Life Story of Rasmus B. Anderson* (1915; repr. Miami, FL: Hardpress, 2010), 92.

6. Anderson and Barton, *Life Story*, 90.

7. Richard Ringmar, "Gluntarne," CD liner notes, *Wennerbergs Gluntarne*, Grammofon AB BIS, Djursholm, Sweden, 1984, 1996: 10.

8. Ringmar, "Gluntarne," 10.

9. *Encyclopaedia Britannica*, 10th ed., s.v. "Wennerberg, Gunnar."

10. Michael Roberts, *From Oxenstierna to Charles XII: Four Studies* (Cambridge: Cambridge University Press, 1991), 59.

11. Birgitta Steene in Warme, *History of Swedish Literature*, 228.

12. Steene in Warme, *History of Swedish Literature*, 227.

13. Anderson and Barton, *Life Story*, 90; Albert O. Barton, "Carl Gustaf Mellberg: Koshkonong Pioneer," *Wisconsin Magazine of History* 29, no. 4 (June 1946): 410.

14. The American poet Henry Wadsworth Longfellow (1807–1882), on a European tour, traveled to Sweden to meet Tegnér; a prose version of the saga appeared in Longfellow's first book in 1839.

15. Nolin in Warme, *History of Swedish Literature*, 189

16. Anderson and Barton, *Life Story*, 92. Gustaf Unonius, in his two-volume memoir, refers to Tegnér and Geijer in such a way that tells us his friends and neighbors understood his references perfectly. *A Pioneer in Northwest America, 1841–1858: The Memoirs of Gustaf Unonius*, vol. 1, trans. Jonas Oscar Backlund, ed. Nils William Olsson (Minneapolis: Swedish Pioneer Historical Society and University of Minnesota Press, 1950).

17. Wheeler, quoted in Edward Lee Greene, "Sketch of the Life of Thure Kumlien, A. M.," in *Pittonia: A Series of Papers Relating to Botany and Botanists, 1887–1889*, vol. 1 (Berkeley, CA: n.p.), 258. Although the rhymed couplets may seem forced to modern ears, there is something sparkling and exciting and sweet in the poem that Thure carried across the Atlantic. Here are a few stanzas from the beginning. Canto I introduces the young Frithiof and Ingeborg, children of King Bele and Thorsten Vikingson, growing up as foster children of Hilding:

> There grew, in Hilding's garden fair,
> Two plants beneath his fostering care;
> Such plants the North had never seen;
> How gloriously they deck the green!
>
> One like the oak-tree soars on high,
> Whose trunk all proudly greets the sky;
> While bending still, by winds caress'd
> Its branches wave like warrior's crest.
>
> The other blossoms like the rose,
> Ere yet the vernal suns disclose
> The charms that in the chalice dawn,
> Though winter hath its breath withdrawn.

Canto II introduces the old men, King Bele and Thorsten Vikingson, friends since their youth, as they consider their coming death. The two sons of the king are brought before him with Frithiof:

"My children," said the aged king, "I feel th' approach of death;
In concord govern ye the land, when I'm deprived of breath:
For, e'en as rings support the lance, concord upholds the throne;
If from the lance you take the rings, its strength at once is gone.

"Let force remain as sentinel to watch the country's door,
But to fair peace an altar raise, and still her smile adore:
For to defend, not injure man, should be the weapon's lot;
And let your shield be ever placed before the peasant's cot.

"Tis but the man insane that seeks t' oppress his native land;
What can a monarch e'er achieve without his people's hand?
The foliage of the glorious oak must wither and decay,
If the bold trunk no life receives from hard and barren clay."

18. Carl Frängsmyr, Magdalena Hydman, and Ragnar Insulander, "Linnaeus as a Priest of Nature," Linné on line, Uppsala University, www2.linnaeus .uu.se/online/history/prasten.html.
19. Frängsmyr, Hydman, and Insulander, "Linnaeus as a Priest of Nature."
20. Both gardens are now restored and open to the public.
21. Thure Kumlien to Frithiof Kumlien, Summer 1877, typed translation, Albion Academy Historical Society, Kumlien document #396. Von Yhlen (spelled von Uhlm sometimes by Kumlien), who later worked in the water fisheries of the west coast of Sweden, drew illustrations for Adolf Erik Nordenskiöld's account of his trip to the Arctic. Von Yhlen may have sent Kumlien a copy of this book.
22. "It has given me great pleasure trying to procure data concerning Thure Kumlien, whom, I very well remember, since the beginning of the 1840's, visited in my parents' home." Theodor Fries, son of Elias Fries, to A. O. Linder, quoted in Angie Kumlien Main, "Thure Kumlien, Koshkonong Naturalist," *Wisconsin Magazine of History* 27, no. 1 (September 1943): 24.
23. Olof Rudbeck the Younger, quoted in James Larson, review of *Book of Birds: A Facsimile of the Original Watercolors of Olof Rudbeck the Younger in the Leufsta Collection in Uppsala University Library*, by Olof Rudbeck and Björn Löwendahl, *History of Biology Reviews—Isis* 79, no. 2 (1988): 297, www .journals.uchicago.edu/doi/abs/10.1086/354712?journalCode=isis.
24. Rudbeck quoted in Larson, 297.
25. Accounts list differing numbers of disciples; some writers list seventeen, others as many as twenty-two; some say eight men died before returning, others seven.

26. A quote from John Ellis to Carl Linnaeus in "Joseph Banks 1743–1820," PlantExplorers.com, www.plantexplorers.com/explorers/biographies /banks/joseph-banks-01.htm.

27. Peter Kalm, *Travels into North America*, trans. John Reinhold Forster (Barre, MA: Imprint Society, 1972), 14.

28. Kalm, *Travels*, 24.

29. Quoted in Robin T. Reid, "The Story of Bartram's Garden," *Smithsonian Magazine*, April 12, 2010, www.smithsonianmag.com/history/the-story -of-bartrams-garden-13572809.

30. Kalm, *Travels*, xviii.

31. Edmund Berkeley and Dorothy Smith Berkeley, *The Life and Travels of John Bartram: From Lake Ontario to the River St. John* (Tallahassee: University Presses of Florida, 1982), 61–63.

32. Pehr Kalm to Linnaeus, quoted in Paula Ivaska Robbins, *The Travels of Peter Kalm: Finnish-Swedish Naturalist, Through Colonial North America, 1748– 1751* (Fleishmans, NY: Purple Mountain Press, 2007), 52.

33. Kalm, *Travels*, 44.

34. Kalm, *Travels*, 237.

35. Kalm, *Travels*, 112.

36. Kalm, *Travels*, 113.

37. Johan Carl Hellberg, *Ur minnet och dagboken om mina samtida: personer och händelser efter 1815 inom och utom fäderneslandet* [From the memory and diary of my contemporaries: Persons and events after 1815 inside and out- side the fatherland], (Stockholm: Iwar Hæggströms boktryckere/Bonnier, 1870); Birger Jarl, "Ångbåtarna på Uppsala, En översikt över seglationerna före 1890" in *Uppland, Upplands Fornminnesförenings Årsbok 1946* (Uppsala, Sweden: 1947): 130–31.

Chapter 3

1. Gustaf Unonius, *A Pioneer in Northwest America, 1841–1858: The Memoirs of Gustaf Unonius*, vol. 1, trans. Jonas Oscar Backlund, ed. Nils William Olsson (Minneapolis: Swedish Pioneer Historical Society and University of Minnesota Press, 1950), 5.

2. Nils William Olsson, "Gustaf Unonius, 1841–1991," *Swedish-American His- torical Quarterly* 43, no. 2 (April 1992): 70.

3. Unonius, *A Pioneer*, vol. 1, 407.

4. Unonius, *A Pioneer*, vol. 1, 6.

5. Unonius, *A Pioneer*, vol. 1, 6.

6. Unonius, *A Pioneer*, vol. 1, 6.

7. Lars Ljungmark, *Swedish Exodus* (Carbondale: Southern Illinois University Press, 1979), 11.

8. Esaias Tegnér, quoted in H. Arnold Barton, ed., *Letters from the Promised Land: Swedes in America, 1840–1914* (Minneapolis: University of Minnesota Press, 1975, 2000), 9.

9. Barton, *Letters from the Promised Land*, 13.

10. Unonius, *A Pioneer*, vol. 1, 7–8.

11. George M. Stephenson, trans. and ed., assisted by Olga Wold Hansen, *Letters Relating to Gustaf Unonius and the Early Swedish Settlers in Wisconsin* (Rock Island, IL: Augustana Historical Society Publications, 1937), 8.

12. Unonius, *A Pioneer*, vol. 1, 10.

13. Unonius, *A Pioneer*, vol. 1, 11.

14. Unonius, *A Pioneer*, vol. 1, 11.

15. Unonius, *A Pioneer*, vol. 1, 40.

16. Albert O. Barton, "Carl Gustaf Mellberg: Koshkonong Pioneer," *Wisconsin Magazine of History* 29, no. 4 (June 1946): 407–22.

17. Stephenson and Hanson, *Letters Relating to Gustaf Unonius*, 40.

18. Stephenson and Hanson, *Letters Relating to Gustaf Unonius*, 40.

19. Stephenson and Hanson, *Letters Relating to Gustaf Unonius*, 43.

20. George M. Stephenson, "Historical Introduction" in *Letters Relating to Gustaf Unonius*, 10.

21. Stephenson and Hanson, *Letters Relating to Gustaf Unonius*, 42–43.

22. Stephenson and Hanson, *Letters Relating to Gustaf Unonius*, 53–64, 93–112.

23. Stephenson and Hanson, *Letters Relating to Gustaf Unonius*, 84.

24. Unonius was probably referring to what we would now call sharp-tailed grouse and greater prairie-chickens, respectively.

25. Stephenson and Hanson, *Letters Relating to Gustaf Unonius*, 88–89.

26. Carl Gustaf Löwenhielm, handwritten manuscript, Archive at the Royal Swedish Academy of Sciences (Kungliga Vetenskapsakademien, KVA), 1–2.

27. In looking into Carl Gustaf Löwenhielm, who later loaned Kumlien the money to go to North America, Lena Peterson Engseth, researcher and translator, and Maria Asp, archivist, located in a file on Löwenhielm at the Royal Swedish Academy of Sciences a draft in Löwenhielm's hand of a report of Kumlien's 1842 trip to Gotland (hereafter cited as Löwenhielm manuscript). Kumlien and/or Cornell must have handed notes and field books over to Löwenhielm, who planned to write up and publish them, as he did on a trip he took the next year to the north of Sweden. The narrative of this Gotland trip had been lost for 176 years. All we knew was that

Kumlien had traveled to one of the Baltic islands in 1842. The original of Kumlien's field book has not been found.

28. Thure Kumlien, Work Journal, 1844, p. 4 (typescript by Albert O. Barton, Wisconsin Historical Society Archives).

29. Löwenhielm manuscript, 3.

30. Löwenhielm manuscript, 3.

31. Löwenhielm manuscript, 3.

32. *Sweriges och Norriges Calender för året 1845* utgifven efter Kongl. Maj:ts nådigste förordnande, af dess wetenskaps-academi (Stockholm: Norstedts & Söner), 479. Article originally published February 15, 1932, in the local newspaper *Gotlands Allehanda* (N:r 37) based on a publication from 1851, "Några ord om Wisby fornlemningar utgifna såsom upplysningar till kartan över Wisby" by O. J. Sjöberg, printed at Theodor Norby, Visby 1851, http://gutarforr.tingstade.com/1932/1042-19320215-en-vagledning-for -turister. Bohman, Lennart, Ett landsorts läroverk. Studier kring Visby Gymnasium 1821–1971, 1971, p. 135–36.

33. Löwenhielm manuscript, 3.

34. Löwenhielm manuscript, 4.

35. Löwenhielm manuscript, 6.

36. Löwenhielm manuscript, 3.

37. Löwenhielm manuscript, 4.

38. Löwenhielm manuscript, 5.

39. Löwenhielm manuscript, 5.

40. Carl Linneaus quoted in Marita Jonsson, photography by Marita and Helga Jonsson, *Linnaeus in Gotland: From the Diary at the Linnean Society, London, to Present-Day Gotland* (Burgsvik: Gotlandsboken, 2007), 36.

41. Linneaus quoted in Jonsson, *Linnaeus in Gotland*, 34.

42. Löwenhielm manuscript, 2.

43. Thure Kumlien to "Dr. J. Twombly, Prest State University of Wis. Madison," undated (between 1871 and 1874), letter draft, TK papers, box 1, folder 3.

44. Löwenhielm manuscript, 6.

45. Kumlien would later encounter a subspecies of this same bird as a common nester at Koshkonong.

46. Löwenhielm manuscript, 6.

47. Arthur Cleveland Bent, *Life Histories of North American Gulls and Terns: Order Longipennes* (Washington, DC: Smithsonian Institution, United States National Museum, 1921), 297.

48. Löwenhielm manuscript, 6.

49. Löwenhielm manuscript, 6.

50. Löwenhielm manuscript, 8.

51. Löwenhielm manuscript, 8.

52. R. Tod Highsmith notes, "The 'common guillemot' mentioned repeatedly in this paragraph is what today in North America we would call the common murre (*Uria aalge*), as distinct from the black guillemot (*Cepphus grille*)."

53. Löwenhielm manuscript, 7.

54. Löwenhielm manuscript, 7.

55. Löwenhielm manuscript, 8.

56. Löwenhielm manuscript, 8.

57. Löwenhielm manuscript, 8.

58. K. Johansson, "Hufvuddragen af Gotlands växttopografi och växtgeografi, grundade på en kritisk behandling av dess kärlväxtflora," Kongl. Svenska Vetenskaps-akademiens Handlingar Ny följd. 29:e bandet ["The Hufvuddragens of Gotland's plant topography and plant geography, was based on a critical treatment of its vascular plant flora," Congl. Swedish Academy of Sciences Actions New sequence. 29th band"], Norstedt & Söner, Stockholm 1896–97.

59. Löwenhielm manuscript, 9.

60. Löwenhielm manuscript, 10.

61. Löwenhielm manuscript, 10.

62. Löwenhielm manuscript, 10.

63. Kevin Winker, "Bird Collections: Development and Use of a Scientific Resource," *The Auk* 122, no. 3 (July 2005): 966.

64. Winker, "Bird Collections," 966.

65. *Svenskt Porträttgalleri* (Swedish Portrait Gallery), digital version, Project Runeberg; Church Records, Steneby Parish, Älvsborgs län *Fregatten Eugenies resa omkring jorden åren 1851–1853*, 2 volumes, published by the Swedish Academy of Sciences, 1855 and 1857, digital version, Project Runeberg.

66. Carl Gustaf Löwenhielm, *Journal och anteckningar i Zoologi och Jagt m.m. under en resa i Luleå Lappmark m.fl. st. Sommaren 1843* [Journal and notes in Zoologi and hunt etc during a trip to Luleå Lappmark and other places. The summer of 1843]. Handwritten manuscript in the Archive at Kungliga Vetenskapsakademien, KVA (Royal Swedish Academy of Sciences), 1.

67. Löwenhielm, *Journal och anteckningar.*

Chapter 4

1. Angie Kumlien Main, "Thure Kumlien, Koshkonong Naturalist," *Wisconsin Magazine of History* 27, no. 1 (September 1943), 24–25.

2. Gun Björkman, "Soldater och dragoner i socknarna Håtuna och Håbo-Tibble i Upplands-Bro kommun," https://docplayer.se/17066177-Soldater

-och-dragoner-i-socknarna-hatuna-och-habo-tibble-i-upplands-bro
-kommun.html.

3. E. G. Trotzig, "Thure Kumlien, Pioneer Naturalist," *Swedish-American Historical Quarterly* 30, no. 3 (July 1979): 197. Trotzig's aunt was Maria Josefina Amalia Kumlien, Thure's sister.

4. "Sweden: History & Background," https://education.stateuniversity.com /pages/1459/Sweden-HISTORY-BACKGROUND.html; "Sweden: Educational System—Overview," https://education.stateuniversity.com/pages /1461/Sweden-EDUCATIONAL-SYSTEM-OVERVIEW.html.

5. *ArkivDigital*, Husförhörslängder for the following parishes: Håbo-Tibble 1824–1833, AI:9, p. 129, p. 98; Håtuna 1822–34, AI:11, p. 95; Sigtuna 1822– 29, AI:6a, p. 11; Husby-Ärlinghundra 1822–28 AI:6, p. 1; Odensala 1826–30 AI:9, p. 86, 1831–1835, AI:10, p. 115; Börje 1820–1834, AI:5a.

6. Church records for Tibble Parish: Household Records "Husförhörslängd" 1824–33 AI:9, p. 129, 1833–1838 (No. 110), 1838–1843 (No. 109), 1854–1858 AI:14, p. 172. Death records, CI:4, 1822–1862, picture 16. Estate Inventories, Håbo Häradsrätt, Uppsala län.

7. Most of this and the following information on Gustaf Mellberg in Sweden comes from Steig-Erland Dagman, "Gustaf Mellberg: From Swedish Academician to American Farmer," *Swedish American Genealogist* 4, no. 4 (1983): 161–69. Dagman is a kinsman of Mellberg.

8. Axel Friman, "Emigrant Sailings Across the Atlantic Around the Year 1840," *Swedish Pioneer Quarterly* 21, no. 3 (1970): 141.

9. Friman, "Emigrant Sailings," 149.

10. Nils William Olsson, *Swedish Passenger Arrivals in New York 1820–1850* (Chicago: Swedish Pioneer Historical Society, 1967), 44–45.

11. Olsson, *Swedish Passenger Arrivals*, 36.

12. Friman, "Emigrant Sailings," 151.

13. Quoted in Friman, "Emigrant Sailings," 150.

14. Johan Olof Liedberg, "Johan Olof Liedberg, His Memoirs," trans. Selma Jacobson, *Swedish Pioneer History Quarterly* 23, no. 4 (1972): 222.

15. Angie Kumlien Main, "Studies in Ornithology at Lake Koshkonong and Vicinity by Thure Kumlien from 1843 to July 1850," *Transactions of the Wisconsin Academy of Sciences, Arts and Letters* 37 (1945): 91. Main had this and the following documents in her possession in 1945, but they are now lost. Based on papers in Thure Kumlien's handwriting written in Swedish.

16. Main, "Koshkonong Naturalist," 26.

17. The details of the travels of the Kumlien-Wallberg-Mellberg party come from a reconstruction of those days by Stieg-Erland Dagman, who referred to census lists and newspaper articles in "the Stockholm City Archives;

The University of Uppsala Library; as well as the records of Gota Canal and Trollhatte Canal, the Stockholm newspaper *Aftonbladet* and the Gothenberg newspaper *Goteborgs Hanclels-och Sjofartstidning.*" From the editor's note (p. 69) to an article by Stieg-Erland Dagman, "The Emigrant's Departure," in *Swedish American Genealogist* 6, no. 2 (1986): 69–72.

18. We don't know how the Wallbergs paid for their passage. It may be that Kumlien paid for them—also with money borrowed from Löwenhielm.

19. Dagman, "Gustaf Mellberg," 70.

20. Dagman, "Gustaf Mellberg," 71.

21. Dagman, "Gustaf Mellberg," 71.

22. Dagman, "Gustaf Mellberg," 71.

23. Liedberg, "His Memoirs," 226.

24. Dagman, "Gustaf Mellberg," 71–72.

25. Liedberg, "His Memoirs," 227.

26. Liedberg, "His Memoirs," 227.

27. Liedberg, "His Memoirs," 226.

28. Friman, "Emigrant Sailings," 145.

29. Friman, "Emigrant Sailings," 146.

30. Friman, "Emigrant Sailings," 145.

31. Friman, "Emigrant Sailings," 146.

32. Dagman, "Gustaf Mellberg," 72.

33. Liedberg, "His Memoirs," 226.

34. Friman, "Emigrant Sailings," 151.

35. Liedberg, "His Memoirs," 226.

36. Liedberg, "His Memoirs," 226–27.

37. Liedberg, "His Memoirs," 227.

38. Liedberg, "His Memoirs," 227.

39. Liedberg, "His Memoirs," 227.

40. Main, "Koshkonong Naturalist," 26.

41. Liedberg, "His Memoirs," 228.

42. "Out of sight of land, the passengers realized they had quite literally entrusted their lives to Nissen. So writing the poem might represent a primitive incantation of a sort, subtle magic—as if by honoring Nissen it might help ensure continued safe passage." Lena Peterson Engseth, email to author, July 31, 2018.

43. Engseth, email to author, July 31, 2018.

44. Liedberg, "His Memoirs," 228.

45. Liedberg, "His Memoirs," 228.

46. Main, "Koshkonong Naturalist," 26.

47. Liedberg, "His Memoirs," 228.

48. Liedberg, "His Memoirs," 228.

49. Liedberg, "His Memoirs," 228.

50. Though we don't know how much time the next two legs of the trip to Milwaukee took the group, we know they were in Milwaukee on August 28 when several of them indicated their intentions to become US citizens. "The fare from Schenectady to Buffalo was $7.75 with board and $5.75 without." Ronald E. Shaw, *Canals for a Nation: The Canal Era in the United States 1790–1860* (Lexington: University Press of Kentucky, 1990), 48.

51. Liedberg, "His Memoirs," 228.

52. Liedberg, "His Memoirs," 228–29.

53. Liedberg, "His Memoirs," 229.

54. Liedberg, "His Memoirs," 229.

55. Liedberg, "His Memoirs," 229.

56. Liedberg, "His Memoirs," 229.

Chapter 5

1. Albert O. Barton, "Carl Gustaf Mellberg, Koshkonong Pioneer," *Wisconsin Magazine of History* 29, no. 4 (1946): 407.

2. Angie Kumlien Main, "Thure Kumlien, Koshkonong Naturalist," *Wisconsin Magazine of History* 27, no. 1 (September 1943): 26.

3. Johan Olof Liedberg, "Johan Olof Liedberg, His Memoirs," trans. Selma Jacobson, *Swedish Pioneer History Quarterly* 23, no. 4 (1972): 230.

4. Their daughter, Anna, born 1827, would marry Samuel Downing; Josephina, born in 1830, later married Carl Gustaf Hammarquist. Casimir, born in 1833, would later marry Mellberg's daughter Amalia. One-year-old twins Carolina and Conrad did not survive the journey from Milwaukee to Koshkonong.

5. Now Clybourn Street.

6. John G. Gregory, *History of Milwaukee Wisconsin*, vol. 1 (Milwaukee: S. J. Clarke Publishing Company, 1931), 317.

7. Margaret Fuller, *Summer on the Lakes, in 1843* (Boston: Charles C. Little and James Brown, 1844), 110.

8. Fuller, *Summer on the Lakes*, 113.

9. Dr. O. P. Wolcott in 1843, quoted in Gregory, *History of Milwaukee*, 205.

10. Gustaf Unonius, *A Pioneer in Northwest America, 1841–1858: The Memoirs of Gustaf Unonius*, vol. 1, trans. Jonas Oscar Backlund, ed. Nils William Olsson (Minneapolis: Swedish Pioneer Historical Society and University of Minnesota Press, 1950), 153.

11. Unonius, *A Pioneer*, 152–53.

12. Fuller, *Summer on the Lakes*, 109–10.

13. Fuller, *Summer on the Lakes*, 110.

14. Fuller, *Summer on the Lakes*, 110–11.

15. Fuller, *Summer on the Lakes*, 112.

16. WHI Lapham papers, Julia Ann Lapham manuscript, 555, October 12, 1843, Wisconsin Historical Society.

17. Lange, born in Göteborg, Sweden, in 1811, was a sailor at the age of thirteen and was said to have been the first Swede in both Chicago and Wisconsin. Ernst W Olson, Anders Schön, and Martin J Engberg, *History of the Swedes of Illinois*, vol. 1, (Chicago: The Engberg-Holmberg Publishing Company, 1908), 182–84. In Milwaukee, after a short and unhappy stay in Chicago, Lange was working for L. J. Farwell in 1841 when a Norwegian waitress, he says, "Gunnhilda," connected Swede Unonius with Swede Lange (Frederick Hale, *Swedes in Wisconsin*, 2nd ed. [Madison, WI: Wisconsin Historical Society Press, 2002], 6). Lange went with Unonius to Pine Lake to smooth his way and "offered his assistance in claiming suitable land." He would serve the same function in 1850 for the famous Swedish writer Fredrika Bremer when he squired her around Milwaukee and Pine Lake (Bremer on Lange and Milwaukee quoted in Hale, 50–51). Even after he moved permanently back to Chicago, Lange spent much of his life aiding Swedish immigrants and forming Swedish societies. Lange's name appears in Kumlien's journal, March 10, 1844: "Received letter from Lange relative to claim." Lange died in Chicago in 1893.

18. Main, "Koshkonong Naturalist," 28.

19. Edward D. Holton in Gregory, *History of Milwaukee*, 280. In those days, there were three routes they could have taken from Milwaukee to the southwest: one through Big Bend and East Troy toward Janesville; one through Waukesha to the Rock River, then to Genesee, Palmyra, and Whitewater where they connected a road to Jefferson and Fort Atkinson; and one through Muskego to the Fox River Waterford and Wilmot.

20. Unonius, *A Pioneer*, 154.

21. Main, "Koshkonong Naturalist," 30.

22. Increase A. Lapham, *Wisconsin, Its Geography and Topography* (1846; repr. Salem, NH: Ayer Company Publishers, 1999), 143.

23. Increase Allen Lapham in Julia Ann Lapham typescript, Increase Lapham Archives, Wisconsin Historical Society, Madison, Wisconsin, p. 489, description of Milwaukee area for Silliman's journal.

24. Lucien B. Caswell quoted in Hugh Highsmith, *The Mounds of Koshkonong and Rock River: A History of Ancient Indian Earthworks in Wisconsin* (Fort Atkinson, WI: Fort Atkinson Historical Society, Highsmith Press, 1997), 10–11.

25. Liedberg, "His Memoirs," 229. Most accounts say one night on the road; Liedberg remembers three.

26. Main, "Koshkonong Naturalist," 29–30.

27. Main, "Koshkonong Naturalist," 29. Reuterskiold bought and paid for this land on October 27, 1843.

28. A. B. Stout and H. L. Skavlem, "The Archaeology of the Lake Koshkonong Region," *The Wisconsin Archeologist* 7 (April–June 1908), reprinted in Highsmith, *The Mounds of Koshkonong*, 197.

29. Stout and. Skavlem, "Archaeology," 197.

30. Lucien Caswell quoted in Highsmith, *The Mounds of Koshkonong*, 8–9.

31. Main, "Koshkonong Naturalist," 29.

32. Main, "Koshkonong Naturalist," 29.

33. Liedberg, "His Memoirs," 230.

34. "In the Land Patent Book in the courthouse at Jefferson, I find that Charles E. A. Reuterskiold bought and paid for 320 acres of land, October 27, 1843. . . . This date makes it seem that the men might have gone to Koshkonong early in September and returned to Milwaukee in the latter part of October," in Main, "Koshkonong Naturalist," 29.

35. Main, "Koshkonong Naturalist," 31.

36. Liedberg, "His Memoirs," 230.

37. Lawrence Martin, *The Physical Geography of Wisconsin* (Madison: The University of Wisconsin Press, 1965), 214.

38. Main, "Koshkonong Naturalist," 30.

39. Mentioned by Pehr Kalm in *Travels into North America* (Barre, MA: Imprint Society, 1972), 375, 386. Kalm saw the *Zizania aquatica* near Lake Champlain and near the St. Lawrence River.

40. Kumlien arrived between the similar descriptions of the lake by Lucien Caswell in 1836 and those of 1850 by Lapham, so we can be certain that Kumlien saw what they saw.

41. Arthur Cleveland Bent, *Life Histories of North American Gulls and Terns: Order Longipennes* (Washington, DC: Smithsonian Institution, United States National Museum, 1921), 297.

Chapter 6

1. The Downings' log cabin was later enclosed in a larger house built by H. L. Skavlem; it is now the Carcajou Hunt Club. There is still a grove of bur oaks on the site.

2. The journal was translated by Albert O. Barton, the son of Norwegian immigrants who lived on the Koshkonong Prairie. The original and Barton's

typescript are at the Wisconsin Historical Society Archives. All quotations are from Barton's typescript. The parentheses are Kumlien's.

3. Abbreviation for "same as above."

4. Kumlien, Work Journal, March 4, 1844, 1 (typescript by Albert O. Barton, Wisconsin Historical Society Archives, Madison).

5. Kumlien, Work Journal, March 22, 1844, 1.

6. Bjorkander was at the time either at Pine Lake or on a trip to New Orleans with Carl Hammarquist. In September 1846, Hammarquist moved to Koshkonong with his wife, Josephine Reuterskiold, who came on the *Svea* with her parents, Carl and Maria Reuterskiold.

7. Hjalmar Rued Holand, *History of the Norwegian Settlements: A translated and expanded version of the 1908 De Norske Settlementers Historie and the 1930 Den Siste Folkevandring Sagastubber fra Nybyggerlivet i Amerika* (Decorah, IA: Anundsen Publishing Company), 65.

8. Rasmus Bjorn Anderson and Albert O. Barton, *Life Story of Rasmus B. Anderson* (1915; repr. Miami, FL: Hardpress, 2010), 91.

9. Anderson and Barton, *Life Story*, 5. The camp was in the "west half of the southeast quarter of section 19 northeast corner in Albion Township, Dane county" twelve miles from Milton.

10. Kumlien, Work Journal, March 20, 1844, p. 2.

11. George W. Ogden, quoted in Hugh Highsmith, *The Mounds of Koshkonong and Rock River: A History of Ancient Indian Earthworks in Wisconsin* (Fort Atkinson, WI: Fort Atkinson Historical Society, Highsmith Press, 1997): 31–32.

12. Increase A. Lapham, *Antiquities of Wisconsin: As Surveyed and Described* (1855; repr. Madison: University of Wisconsin Press, 2001), 35.

13. In 1922, Theodore Kumlien, Thure's second son, who had taken over his father's farm after his death, sold a collection of Indian artifacts to the Logan Museum at Beloit. "Archeological Notes," *Wisconsin Archeologist* I, no. 3 (1922): 112.

14. Bo Wallin, *A Richer Forest* (Stockholm: National Board of Forestry of Sweden, 1992), 44.

15. Called the Royal Mounds for years, the first of the mounds was excavated in 1847. In the nineteenth century, when Thure was at the university, they were called after important kings of Sweden: *Aun's Mound*, *Adil's Mound*, and *Egil's Mound*. Today they are called the *Eastern Mound*, *Middle Mound*, and *Western Mound*. Dated to the fifth and sixth centuries, they are Sweden's oldest national symbols, depicted on the covers of books about the Swedish national identity. In the sixth century, Gamla Uppsala was the location of royal burials.

16. Though one effigy mound at the north end was partially destroyed when a section of road was cut through, all the owners of the property since Thure Kumlien have left the mounds undisturbed.

17. The Ho-Chunk had signed a treaty ceding the rights to their land north of the Rock River and south of the Wisconsin River in 1829. After the tragic miscommunications and manipulations that resulted in a massacre of Sauk people in the Black Hawk War, the Potawatomi and Ojibwe Nations sold vast acres of their land, and the Ho-Chunk ceded their land south and east of the Wisconsin River and the Fox River in 1832.

18. Amy Lonetree, "Visualizing Native Survivance: Encounters with My Ho-Chunk Ancestors in the Family Photographs of Charles Van Schaick," in *People of the Big Voice: Photographs of Ho-Chunk Families by Charles Van Schaick, 1879–1942* (Madison: Wisconsin Historical Society Press, 2011), 19.

19. Juliette M. Kinzie, *Wau-Bun: The "Early Day" in the North-West* (1856; repr. Urbana: University of Illinois Press, 1992), 174–75. In 1831, Juliette Kinzie was a fine writer and the wife of John Kinzie, Indian agent at Fort Winnebago at Portage, Wisconsin. In a well-known account following the Fort Dearborn massacre in Chicago, Juliette wrote of their travel from Chicago to Portage. On the way, they visited an Indian village at Koshkonong, and from her writing, we get a glimpse of a peaceful and prosperous village. After "a ride of five or six miles through the beautiful oak openings," the Kinzie party came to the place where the Rock River opens to Lake Koshkonong. Here they found "a collection of neat bark wigwams, with extensive fields on each side of corn, beans, and squashes, recently planted, but already giving promise of a fine crop. In front [of the village] was the broad blue lake, the shores of which, to the south, were open and marshy, but near the village, and stretching far away to the north, were bordered by fine lofty trees."

Chapter 7

1. Gustaf Unonius, *A Pioneer in Northwest America, 1841–1858: The Memoirs of Gustaf Unonius*, vol. 1, trans. Jonas Oscar Backlund, ed. Nils William Olsson (Minneapolis: Swedish Pioneer Historical Society and University of Minnesota Press, 1950), 304–5. The officer, Johan Frederic Polycarpus von Schniedau and his wife, Carolina Elizabeth Jacobsson, left the Pine Lake settlement for Chicago where he became one of the first daguerreotypists in Chicago. The brother-in-law who came with them gave up farming to become a merchant with a small country store, then sold everything and returned to Sweden.

2. Frederick Israel Unonius was born November 27, 1842, and died May 11, 1846. Their second son, Lloyd Gustaf Breck Unonius, was born August 14, 1844. Lloyd Gustaf apparently returned to Sweden with his parents in 1858 but came back to America around the time of the Civil War, settling in Chicago and marrying. His granddaughter reported that he walked away from his home one Sunday morning and was never seen again. Gustaf Unonius, *A Pioneer in Northwest America, 1841–1858: The Memoirs of Gustaf Unonius*, vol. 2, trans. Jonas Oscar Backlund, ed. Nils William Olsson (Minneapolis: Swedish Pioneer Historical Society and University of Minnesota Press, 1960), 329.

3. At the time Agusta was born, Thure and Christine had been married for about seven months—since September 5, 1843. Christine likely became pregnant when they were on board the *Svea*.

4. No records survive to tell us which museums he sent boxes to at this time, though one was likely the museum of the Skara Gymnasium.

5. Thure Kumlien, Work Journal, October 15, 1845, p. 17 (typescript by Albert O. Barton, Wisconsin Historical Society Archives, Madison).

6. Kumlien, Work Journal, February 23, 1845, 9.

7. Kumlien, Work Journal, March 30, 1845, 10. Thure leaves blank the names of the museums.

8. Kumlien, Work Journal, July 1845, 14.

9. Kumlien, Work Journal, September 1–6, 1845, 15.

10. Kumlien, Work Journal, October 1, 8, 1845, 16.

11. Kumlien, Work Journal, October 16, 1845, 17.

12. Kumlien, Work Journal, October 23, 1845, 17.

13. Kumlien, Work Journal, October 24, 1845, 17–18.

14. Christine, Thure, and their son Frithiof would be buried in this same plot.

15. Kumlien, Work Journal, October 26, 1845, 18.

16. Kumlien, Work Journal, January 1, 1846, 23.

17. Kumlien, Work Journal, January 6, 1846, 23

18. Unonius, *A Pioneer*, vol. 2, 105.

19. Unonius, *A Pioneer*, vol. 2, 105.

20. Unonius, *A Pioneer*, vol. 2, 106.

21. Kumlien, Work Journal, January 29, 1846, 24.

22. Kumlien, Work Journal, February 11, 1846, 24.

23. Kumlien, Work Journal, March 1, 1846, 25.

24. Kumlien, Work Journal, March 31, 1846, 27.

25. Kumlien, Work Journal, April 30, 1846, 28.

26. Kumlien, Work Journal, May 10, 1846, 29.

27. Kumlien, Work Journal, August 20–26, 1846, 32–33.

28. Axel Friman, "The First Parish Records of Gustaf Unonius," *Swedish-American Historical Quarterly* 23, no. 4 (1972): 271.

29. Kumlien, Work Journal, January 3, 7, 10, 16, 1847, 37.

30. Kumlien, Work Journal, February 10–17, 1847, 38.

31. Unonius, *A Pioneer*, vol. 1, 331.

32. Kumlien, Work Journal, April 10, 1846, 27.

33. Angie Kumlien Main, "Thure Kumlien, Koshkonong Naturalist (II)," *Wisconsin Magazine of History* 27, no. 2 (December 1943): 204.

34. Kumlien, Work Journal, April 19, 1849, 64, and November 20, 1849, 70.

35. "I have read that Bjorkander received a fatal wound from the machinery in his sawmill near-by which he then operated. On Sunday, April 4, 1943, Mr. Main and I visited the place where Swen's house formerly stood, in our old field. There is a slight depression which shows where the cellar had been filled up." Angie Kumlien Main, "Thure Kumlien, Koshkonong Naturalist (III)" *Wisconsin Magazine of History* 27, no. 3 (March 1944): 325, footnote.

36. Main, "Koshkonong Naturalist," *Wisconsin Magazine of History* 27, no. 3: 325.

Chapter 8

1. Thure Kumlien, Work Journal, May 25, 1844, 4, 6 (typescript by Albert O. Barton, Wisconsin Historical Society Archives).

2. Benjamin Dole is listed as a head of a household selling dry goods in Buffalo city directories from 1842 through 1849. The only other Doles listed are a public school teacher, John A. Dole, 1844 and 1848–1849, and Thomas D. Dole, a clerk at an elevator company living in a boarding house, 1848–1849. *Cuttings Buffalo New York Directories* (Buffalo, New York: Polk-Clement City Directories, 1842, 1844, 1849), 1849, p. 143.

3. Angie Kumlien Main, "Thure Kumlien, Koshkonong Naturalist (II), *Wisconsin Magazine of History* 27, no. 2 (December 1943): 209.

4. Angie Kumlien Main, "Studies in Ornithology at Lake Koshkonong and Vicinity by Thure Kumlien from 1843 to July 1850," *Transactions of the Wisconsin Academy of Sciences, Arts and Letters* 37 (1945): 101.

5. Main, "Studies in Ornithology," 101.

6. Catesby visited North America in 1712–1714 and 1722–1726. Pehr Kalm visited North America from 1748–1751.

7. Edward H. Burtt Jr. and William E. Davis Jr., *Alexander Wilson: The Scot who Founded American Ornithology* (Cambridge: Belknap Press of Harvard University Press, 2013), 35.

8. Kumlien, Work Journal, May 6, 1848, 51.

9. Kumlien, Work Journal, January 25, 1847, 37.

10. Kumlien, Work Journal, April 10, 1847, 39.
11. Kumlien, Work Journal, April 14, 1847, 39.
12. Kumlien, Work Journal, November 6, 1847, 45. Bell apparently did not answer his letter, which has not survived.
13. Kumlien, Work Journal, May 29, 1848, 51.
14. Kumlien, Work Journal, March 20, 1849, 62.
15. Kumlien, Work Journal, March 22, 1849, 62.
16. Kumlien, Work Journal, September 28 and 29, 1849, 69.
17. Main, "Koshkonong Naturalist (II)," 207; Kumlien, Work Journal, October 31, 1849, 69.
18. TK papers, box 1, folder 1.
19. 1840 and 1846 census quoted in Robert C. Nesbit and William F. Thompson, *Wisconsin, A History* (Madison: University of Wisconsin Press, 1989), 212–13.
20. "Charles Holt Is Veteran Editor," *Janesville Gazette*, September 26, 1907, Wisconsin Local History and Biographical Articles Collection, Wisconsin Historical Society, Madison.
21. Main, "Koshkonong Naturalist (II)," 210.
22. Main, "Koshkonong Naturalist (II)," 210.
23. They met on June 14, 1850. Main, "Studies in Ornithology," 103.
24. Main, "Studies in Ornithology," 106.
25. Increase Allen Lapham in Julia Ann Lapham typescript, Increase Lapham Archives, Wisconsin Historical Society, Madison, Wisconsin, p. 715.
26. Lapham, Journal, 715.
27. Lapham, Journal, 716.

Chapter 9

1. We don't have Thomas Brewer's enclosed letter, but Charles Holt's transmitting letter of July 23, 1850, is quoted in Angie Kumlien Main, "Thure Kumlien, Koshkonong Naturalist (II)," *Wisconsin Magazine of History* 27, no. 2 (December 1943): 210.
2. We have most of the letters Brewer sent to Kumlien and only some of the practice drafts of letters Kumlien sent to Brewer. The draft of the first letter Kumlien apparently sent to Brewer is dated March 30, 1851, and is quoted in Main, "Koshkonong Naturalist (II)," 211–12.
3. Brewer to Kumlien, November 28, 1851, TK papers, box 1, folder 1.
4. Brewer's blackbird, Brewer's duck, Brewer's shrew mole. R. Tod Highsmith notes, "Brewer's duck, named by Audubon, was actually a hybrid between a mallard and gadwall and did not hold its species status for very long."
5. Main, "Koshkonong Naturalist (II)," 211–12.

6. Brewer to Kumlien, July 8, 1851.

7. Brewer to Kumlien, July 8, 1851.

8. R. Tod Highsmith notes that we now refer to many of these birds by different names: sparrow hawk [American kestrel], whippoorwill [Eastern whip-poor-will], chimney swallow [chimney swift], kingbird [Eastern kingbird], white-rumped shrike [loggerhead shrike], great grey shrike [Northern shrike], oriole [either Baltimore oriole or the orchard oriole], hummingbird [ruby-throated hummingbird], pinneated grouse [greater prairie-chicken], whooping crane [it is possible Kumlien meant the much more common sandhill crane], bittern [either American bittern or least bittern], American egret [great egret], upland plover [upland sandpiper], and gull [most likely ring-billed gull].

9. Kumlien to Brewer, March 30, 1851, in Main, "Koshkonong Naturalist (II)," 211–12.

10. Kumlien to Brewer, March 30, 1851, in Main, "Koshkonong Naturalist (II)," 211–12.

11. Kumlien to Brewer, March 30, 1851, in Main, "Koshkonong Naturalist (II)," 211–12.

12. Undated letter fragment, TK papers, box 1, folder 1.

13. Main, "Koshkonong Naturalist (II)," 208.

14. Main, "Koshkonong Naturalist (II)," 208.

15. R. Tod Highsmith notes, "Prior to the recent establishment of a breeding population of whooping cranes at Necedah National Wildlife Refuge, there are no accepted records of nesting by whooping cranes in Wisconsin, nor did Kumlien or his son Ludwig ever report that they found nests. Kumlien likely bought or traded for these eggs from another collector and then passed them on to Brewer."

16. Brewer to Kumlien, November 28, 1851, in Main, "Koshkonong Naturalist (II)," 212–13.

17. Brewer to Kumlien, November 28, 1851, in Main, "Koshkonong Naturalist (II)," 212–13. R. Tod Highsmith notes, "In this sentence, Brewer is referring to the taxonomic confusion in his day as to whether the 'dusky duck' and 'black duck' were separate species or merely subspecies of the black duck (*Anas rubripes*), as we consider them today. Audubon had described the 'dusky duck' (*Anas obscura*) as a separate species, but the ornithological community overruled him and called it a subspecies. Brewer may be asking whether Kumlien can supply him with either or both subspecies of the black duck in addition to the teal."

18. R. Tod Highsmith notes, "Brewer probably described this wren as 'very rare' because it does not occur in New England and he may not have had

many skins of it. The sedge wren would have been common in proper habitat in southern Wisconsin in Kumlien's day. The same is true for yellow-headed blackbird."

19. Brewer to Kumlien, November 28, 1851, in Main, "Koshkonong Naturalist (II)," 212–13.

20. Brewer to Kumlien, November 28, 1851, in Main, "Koshkonong Naturalist (II)," 212–13. R. Tod Highsmith notes, "This suggests that Kumlien was apparently insecure about his shrike identifications (the two species still give many birders trouble today). If it was collected during the nesting season, the bird was almost certainly a loggerhead shrike."

21. Brewer to Kumlien, November 28, 1851, in Main, "Koshkonong Naturalist (II)," 213–14. R. Tod Highsmith notes, "It is not clear here if Brewer means the whooping crane or great egret. Samuel D. Robbins Jr., in his 1991 *Wisconsin Birdlife: Population and Distribution Past and Present*, notes that great egrets (*Ardea alba*) were not uncommonly mistaken for the rarer whooping cranes in nineteenth-century Wisconsin. Brewer's interest in working out the 'life histories' of these three species indicates some confusion on his part."

22. Many of these local and decorative bird mounts by Thure Kumlien can be seen in the Hoard Museum in Fort Atkinson, Wisconsin.

23. Brewer to Kumlien, November 28, 1851, in Main, "Koshkonong Naturalist (II)," 214. R. Tod Highsmith notes, "This paragraph confirms Kumlien's confusion about the different crane species, and confusion among the locals calling the great egret a 'white crane.' As for the lack of 'the crest or tuft,' Kumlien is presumably referring to the bird's nuptial plumes (which would later become in great demand for women's hats), and because those feathers become worn or are molted after the nesting season, it's not surprising he didn't see them in August or September."

24. Brewer to Kumlien, November 28, 1851, in Main, "Koshkonong Naturalist (II)," 213.

25. Undated practice copy, Angie Kumlien Main, "Studies in Ornithology at Lake Koshkonong and Vicinity by Thure Kumlien from 1843 to July 1850," *Transactions of the Wisconsin Academy of Sciences, Arts and Letters* 37 (1945): 91–109, note 11.

26. Main, "Studies in Ornithology," 96–97, note 11.

27. Kumlien to Brewer, January 19, 1852, TK papers, box 1, folder 1.

28. T. M. Brewer, *Wilson's American Ornithology, with Notes by Jardine: To Which Is Added a Synopsis of American Birds, including Those Described by Bonaparte, Audubon, Nuttall, and Richardson* (New York: T. L. Magagnos & Company, 1844), 710.

29. Kumlien to Brewer, March 30, 1851, in Main, "Koshkonong Naturalist (II)," 211–12.

30. R. Tod Highsmith notes, "Yellow-heads are more sensitive to water level than red-wings and like deep water marshes. If water levels fall below their preference, yellow-heads may abandon that marsh for the year. So, Kumlien might have observed them nesting nearby in some years, and not at all in others."

31. Kumlien to Brewer, undated draft, dated 1853 by Angie Kumlien Main.

32. Angie Kumlien Main, "Yellow-headed Blackbirds at Lake Koshkonong and Vicinity," *Wisconsin Academy of Sciences, Arts, and Letters* 23 (1927): 632.

33. Main, "Yellow-headed Blackbirds," 632.

34. Main, "Yellow-headed Blackbirds," 632.

35. Main, "Yellow-headed Blackbirds," 634.

36. Main, "Yellow-headed Blackbirds," 634.

37. Main, "Yellow-headed Blackbirds," 632.

38. Elliott Coues, *Field Ornithology* (Boston: Estes & Lauriat, 1874), 53.

39. As I looked at the papers of Thure Kumlien, I wondered at the number of letters on heavier paper that had been trimmed so that all that remained was the text. Any blank paper had been cut off. I suspect that paper was in such short supply for Kumlien that he cut up letters to make slips for labels for birds he sent out.

40. Kumlien to Brewer, August 25, 1854, in Main, "Koshkonong Naturalist (II)," 218.

41. Main, "Yellow-headed Blackbirds," 634.

42. Main, "Yellow-headed Blackbirds," 634.

43. *Proceedings of the Boston Society of Natural History 1854*, vol. 5 (Boston Society of Natural History, 1854), 11. R. Tod Highsmith notes that we now refer to many of these birds by different names: pigeon hawk [merlin], shore lark [horned lark], ruby-crowned wren [ruby-crowned kinglet], common American gallinule [common moorhen], pinneated grouse [greater prairie-chicken], and American black tern [black tern].

Chapter 10

1. Margaret Welch, *The Book of Nature: Natural History in the United States 1825–1875* (Boston: Northeastern University Press, 1998), 2–3.

2. Kumlien to Brewer, undated draft, dated 1853 by Angie Kumlien Main, TK papers, box 1, folder 1.

3. Kumlien to Brewer, undated draft.

4. Kumlien to Brewer, undated draft.

5. Brewer to Kumlien, January 30, 1854, in Angie Kumlien Main, "Thure Kumlien, Koshkonong Naturalist (II)", *Wisconsin Magazine of History* 27, no. 2 (December 1943): 216.

6. Kumlien to Brewer, February 1854. TK papers, box 1, folder 1.

7. R. Tod Highsmith notes, "Today, most taxonomists consider North American and European black terns to be different subspecies of *Chlidonias niger* (although some consider them separate species). This may explain the differences in the color and size of the terns Kumlien collected at Koshkonong as compared to descriptions in *Scandinavian Fauna*."

8. Arthur Cleveland Bent, *Life Histories of North American Gulls and Terns: Order Longipennes* (Washington, DC: Smithsonian Institution, United States National Museum, 1921), 297.

9. Bent, *Life Histories*, 297.

10. Kumlien to Brewer, August 25, 1854, in Main, "Koshkonong Naturalist (II)," 218.

11. Kumlien to Brewer, August 25, 1854, in Main, "Koshkonong Naturalist (II)," 218.

12. Kumlien to Brewer, August 25, 1854, in Main, "Koshkonong Naturalist (II)," 218.

13. Kumlien to Brewer, August 25, 1854, in Main, "Koshkonong Naturalist (II)," 218.

14. Kumlien to Brewer, August 25, 1854, in Main, "Koshkonong Naturalist (II)," 218.

15. Kumlien to Brewer, August 25, 1854, in Main, "Koshkonong Naturalist (II)," 218.

16. Brewer to Kumlien, September 26,1854, in Main, "Koshkonong Naturalist (II)," 219.

17. Kumlien to Brewer, August 25, 1854, in Main, "Koshkonong Naturalist (II)," 218–19.

18. Kumlien to Brewer, August 25, 1854, in Main, "Koshkonong Naturalist (II)," 218–19.

19. Brewer to Kumlien, November 20, 1854, in Angie Kumlien Main, "Thure Kumlien, Koshkonong Naturalist (III)," *Wisconsin Magazine of History* 27, no. 3 (March 1944): 322.

20. R. Tod Highsmith notes, "Both Audubon and Alexander Wilson made a number of similar errors, which are quite understandable given the pioneering nature of their work and lack of other experts and reference books."

21. John Cassin, "*Vireosylvia philadelphica*, nobis,"*Proceedings of the Academy of Natural Sciences of Philadelphia* V (1851), 153.

22. Brewer to Kumlien, January 11, 1856, TK papers, box 1, folder 1.

23. R. Tod Highsmith notes, "Warbling vireos would, indeed, have been a common bird in the shoreline trees around Lake Koshkonong."

24. Thomas Mayo Brewer, "Paper on Vireosylvia," *Proceedings of the Boston Society of Natural History* VI (1859), 109.

25. Kumlien quoted in Brewer, "Paper on Vireosylvia," 110.

26. Kumlien quoted in Brewer, "Paper on Vireosylvia," 110.

27. Kumlien quoted in Brewer, "Paper on Vireosylvia," 110.

28. Brewer to Kumlien, November 20, 1854.

29. Kumlien quoted in Brewer, "Paper on Vireosylvia," 110–11.

30. Brewer, "Paper on Vireosylvia, 108–11.

31. Brewer, "Paper on Vireosylvia, 109.

32. William Moskoff and Scott K. Robinson, "Philadelphia Vireo," Cornell Lab of Ornithology, Birds of North America, April 21, 2011, https://birdsna.org/Species-Account/bna/species/phivir/introduction.

33. Between 1851 and 1860, Brewer wrote Kumlien at least twenty-one letters.

34. Undated letter [1855?] to Baird, TK papers, box 1, folder 1.

35. Angie Kumlien Main, "Studies in Ornithology at Lake Koshkonong and Vicinity by Thure Kumlien from 1843 to July 1850," *Transactions of the Wisconsin Academy of Sciences, Arts and Letters* 37 (1945), 91.

Chapter 11

1. Kumlien to Elias Fries, August 15, 1859. Letters to Elias Fries from Thure Kumlien [in Letters to Elias Fries. From Scandinavians (Senders). H–K. : manuscript G 70 p fol.], 1859–1863, Uppsala University Library, Manuscript and Music collections, http://urn.kb.se/resolve?urn=urn:nbn:se:alvin:portal:record-158366.

2. Fries was only the seventh director of Hortus Botanicus, which had been founded by Olaf Rudbeck the Elder in 1653. Carl Linnaeus was the most important director, from 1741 to 1778. Carl Peter Thunberg, who collected plants in South Africa and Japan, directed Hortus Botanicus from 1783 to 1828. Fries was the first to print a seed catalogue, and he initiated an official exchange of seeds with far-flung botanical gardens and collectors.

3. Kumlien to Fries, August 15, 1859.

4. Kumlien to Fries, August 15, 1859.

5. Kumlien to Fries, August 15, 1859.

6. Kumlien to Fries, August 15, 1859.

7. Kumlien to Fries, August 15, 1859.

8. Alphonso Wood, *Class-Book of Botany: Being Outlines of the Structure, Physiology and Classification of Plants; with a Flora of All Parts of the United States*

and Canada (Claremont, NH: Manufacturing Co., 1847). Numerous editions followed the 1847 printing.

9. *Flora Scanica* (Uppsala, 1835) and *Lichenographia europaea reformata* (Lund, 1831).

10. Kumlien to Fries, August 15, 1859.

11. Kumlien to Fries, March 27, 1860.

12. Kumlien to Fries, March 27, 1860.

13. Kumlien to Fries, March 27, 1860. Kumlien would ship boxes to the Swedish consul in New York, who then put them on a ship bound for Sweden.

14. Kumlien to Fries, March 27, 1860.

15. Kumlien to Fries, March 27, 1860. *Parmeliaceae* is the largest family of lichen-forming fungi. *Peltigera* is commonly known as dog lichen.

16. Kumlien to Fries, March 27, 1860.

17. Kumlien to Fries, March 27, 1860.

18. Kumlien to Fries, February 28, 1861.

19. Kumlien to Fries, February 28, 1861.

20. Kumlien to Fries, February 28, 1861.

21. Kumlien to Fries, February 28, 1861.

22. Kumlien to Fries, February 28, 1861. Increase A. Lapham, in 1840, collected a *synthyris*, which is in the Milwaukee Public Museum. It is unclear which *synthyris* Kumlien found.

23. Kumlien to Fries, February 28, 1861.

24. Brewer to Kumlien, February 12, 1862, TK papers, box 1, folder 1.

25. Kumlien to Fries, February 28, 1861.

26. Kumlien to Fries, July 16, 1861.

27. Carl Jakob Sundevall (1801–1875) was a Swedish zoologist educated at Lund University. He was professor and "keeper" of vertebrate zoology at the Swedish Museum of Natural History from 1839 to 1871.

28. Auguste Sallé (1832–1896) was a French entomologist and dealer in natural history specimens who had collected in Mexico.

29. Dr. Philo Hoy was an ornithologist in Racine at the time.

30. C. G. Löwenhielm Klockhammar, Örebro län, Sweden, to Kumlien, May 8, 1861, trans. Lena Peterson Engseth, TK papers, box 1, folder 1.

31. Angie Kumlien Main, "Thure Kumlien, Koshkonong Naturalist (III)," *Wisconsin Magazine of History* 27, no. 3 (March 1944): 333–34.

32. Kumlien's specimen of the sedge is at the Milwaukee Public Museum.

33. Main, "Koshkonong Naturalist (III)," 333–34.

34. Kumlien to Brewer August 11, 1861, TK papers, box 1, folder 1.

35. Spencer Fullerton Baird, John Cassin, and George Newbold Lawrence, *The Birds of North America* (Philadelphia: Lippincott, 1860).

36. Kumlien to Brewer, August 11, 1861, TK papers, box 1, folder 1.

37. Kumlien to Cassin, April 3, 1862, TK papers, box 1, folder 1.

38. Kumlien to Fries, January 30, 1862.

39. Kumlien to Fries, January 30, 1862.

40. Quoted in Main, "Koshkonong Naturalist (III)," 327.

41. Main, "Koshkonong Naturalist (III)," 327.

42. Main, "Koshkonong Naturalist (III)," 327.

43. Edward Green, the youngest of the family, joined Company K on August 21, 1862. William and Charles had joined in September 1861, and Mansir in December 1863, according to the regimental roster.

44. William, Charles, and Mansir Green were in Company A; Edward was in Company K.

45. Harley Harris Bartlett, "The Botanical Work of Edward Lee Greene," *Torreya* 16, no. 7 (July 1916): 153.

46. Milo Quaife, "The Panic of 1862 in Wisconsin," *Wisconsin Magazine of History* 4, no. 2 (December 1920): 177.

47. Quaife, "The Panic of 1862," 177.

48. Kumlien to Fries, February 10, 1863.

49. Greene to Kumlien, Dover, Tennessee, September 10, 1862, in Angie Kumlien Main, "Life and Letters of Edward Lee Greene," *Transactions of the Wisconsin Academy of Sciences, Arts and Letters* 24 (1929): 148–49.

50. Willis Linn Jepson, "Edward Lee Greene the Individual," *American Midland Naturalist* 30, no.1 (July 1943): 3–5.

51. Greene to Kumlien, September 10, 1862, in Main, "Life and Letters," 148–49.

52. Greene to Kumlien, Fall 1862, in Main "Life and Letters," 149–50.

53. Greene to Kumlien, November 16, 1862, in Main, "Life and Letters," 151–53.

54. Kumlien to Fries, February 10, 1863.

55. Kumlien to Fries, February 10, 1863.

56. Greene to Kumlien, March 12, 1863, in Main, "Life and Letters," 153–54.

57. Greene to Kumlien, May 6, 1863, in Main, "Life and Letters," 155–56.

58. Kumlien to Fries, June 10, 1863.

59. Thore Magnus Fries did become director of the Uppsala Botanic Garden, but not until 1877.

60. Kumlien to Fries, June 10, 1863.

61. Greene to Kumlien, Fort Donelson, July 15, 1863, in Main, "Life and Letters," 158–60.

62. Greene to Kumlien, Fort Donelson, July 15, 1863, in Main, "Life and Letters," 158–60.

63. Kumlien to Fries, August 25, 1863.

64. Main, "Koshkonong Naturalist (III)," 335.

65. Kumlien to Fries, August 25, 1863.

66. Greene to Kumlien, Huntsville, Alabama, January 4, 1865, in Main, "Life and Letters," 160.

67. Greene to Kumlien, January 4, 1865, in Main, "Life and Letters," 161.

68. Greene to Kumlien, January 4, 1865, in Main, "Life and Letters," 161.

69. Greene to Kumlien, January 4, 1865, in Main, "Life and Letters," 161.

70. Greene to Kumlien, Huntsville, Alabama, March 18, 1865, in Main, "Life and Letters," 163.

71. Greene to Kumlien, Nashville, Tennessee, May 2, 1865, in Main, "Life and Letters," 165.

72. Greene to Kumlien, May 2, 1865, in Main, "Life and Letters," 164.

73. Greene to Kumlien, Huntsville, Alabama, March 18, 1865, in Main, "Life and Letters," 163.

74. Greene to Kumlien, Nashville, Tennessee, May 2, 1865, in Main, "Life and Letters," 165.

75. Edward Lee Greene quoted in Bartlett, "Botanical Work," 152, 154.

Chapter 12

1. Edward A. Samuels to Kumlien, March 20, 1866, TK papers, box 1, folder 1. The full title is *The Ornithology and Oology of New England: Containing full descriptions of the birds of New England and adjoining states and provinces, arranged by a long-approved classification and nomenclature: together with a complete history of their habits, times of arrival and departure: with illustrations of many species of the birds and accurate figures of their eggs* (Boston: Nichols and Noyes, 1867).

2. Samuels to Kumlien, March 20, 1866.

3. J. Q. Emery, "Albion Academy," *Wisconsin Magazine of History* 7, no. 3 (March 1924): 313.

4. Lloyd Hustvedt, *Rasmus Bjorn Anderson: Pioneer Scholar* (Northfield, Minnesota: The Norwegian-American Historical Association, 1966), 53.

5. Hustvedt, *Rasmus Bjorn Anderson*, 19.

6. Svea M. Adolphson, *A History of Albion Academy, 1853–1918* (Beloit, WI: Rock County Rehabilitation Services, 1976), 9–12.

7. Emery, "Albion Academy," 311.

8. Adolphson, *History of Albion Academy*, 23.

9. Founding of Milton Academy, later Milton College, 1844; Carroll College, Waukesha, 1846; Beloit College, 1847; Mount Mary College, Milwaukee, 1850; Ripon College, 1851; Racine College, 1852; Downer College, Milwaukee, 1853; Albion Academy, 1853. Many of those with religious affiliation dropped those connections. Though provision for teachers' colleges, or state normal

schools, was in the 1848 constitution, the Platteville Normal School opened in 1866, Whitewater in 1868, Oshkosh in 1871, and River Falls in 1875, offering two-year certificate degrees and four-year diploma degrees. Free high schools were not provided by law until 1875.

10. Emery, "Albion Academy," 310.

11. Angie Kumlien Main, "Thure Kumlien, Koshkonong Naturalist (III)," *Wisconsin Magazine of History* 27, no. 3 (March 1944): 332.

12. W. C. Whitford to Kumlien, August 27, 1866, TK papers, box 1, folder 1.

13. Rasmus B. Anderson and Albert O. Barton, *Life Story of Rasmus B. Anderson* (1915; repr. Miami, FL: Hardpress, 2010), 88.

14. Anderson and Barton, *Life Story*, 89.

15. Anderson and Barton, *Life Story*, 92.

16. Hustvedt, *Rasmus Bjorn Anderson*, 80. Even to the meticulous Hustvedt, this story is murky.

17. R. Tod Highsmith notes, "This bird very possibly came from Racine, where the first known introduction of the species to Wisconsin occurred in 1869." Samuel D. Robbins Jr., *Wisconsin Birdlife: Population and Distribution Past and Present*, 1991.

18. Rasmus to Abel Anderson, January 21, 1869, quoted in Hustvedt, *Rasmus Bjorn Anderson*, 83.

19. Kumlien's resignation is in the Anderson papers, quoted in Hustvedt, *Rasmus Bjorn Anderson*, 85. Albion Academy was never the same after this, though Anderson's and Kumlien's leaving were only part of the cause of the decline. Changing demographics and increasing competition from other schools weakened Albion's draw and enrollment. Albion was purchased from the Seventh Day Baptists in 1890 and survived in other forms until 1918.

20. Kumlien to Anderson, March 21, 1869, Anderson Papers, Wisconsin Historical Society, Madison (hereafter RBA papers), box 18.

21. Kumlien to Anderson, March 21, 1869, RBA papers, box 18.

22. Kumlien to Anderson, April 7, 1869, RBA papers, box 18.

23. Kumlien to Anderson, April 7, 1869.

24. Kumlien to Anderson, April 7, 1869. Kumlien's emphasis.

25. Kumlien to Anderson, April 7, 1869.

26. Kumlien to Anderson, April 7, 1869.

27. Kumlien to Anderson, April 13, 1869, RBA papers, box 18.

28. Kumlien to Anderson, October 14, 1869, RBA papers, box 18.

29. Kumlien to Anderson, December 30, 1869, RBA papers, box 18.

30. Brewer to Kumlien, February 9, 1871, TK papers, box 1, folder 1.

31. Brewer to Kumlien, February 9, 16, and 25; March 31; April 16, 1871, TK papers, box 1, folder 1.

32. Brewer to Kumlien, November 13, 1871.

33. Brewer to Kumlien, December 7, 1871.

34. Brewer to Kumlien, November 21, 1872.

35. Kumlien to Anderson, January 1, 1872, RBA papers, box 18. We don't know who Lina was.

36. Kumlien to Anderson, January 1, 1872, RBA papers, box 18.

37. Kumlien to John Hanson Twombly, undated, TK papers, box 1, folder 3.

38. Kumlien to Anderson, January 1, 1872.

39. Kumlien to Anderson, January 1, 1872.

40. Kumlien to Anderson, March 30, 1872, RBA papers, box 18.

41. Kumlien to Anderson, March 30, 1872.

42. Kumlien to Anderson, March 30, 1872.

43. Kumlien to Anderson, March 30, 1872.

44. John Oncken, "Cross Country: Tobacco Farming Has a Long History in Wisconsin," *Capital Times*, February 4, 2010.

45. Brewer to Kumlien, January 14, 1873; Brewer to Kumlien, February 1, 1873, TK papers, box 1, folder 1.

46. Ludwig Kumlien and Ned Hollister, *The Birds of Wisconsin*, Bulletin of the Wisconsin Natural History Society, vol. 3, no. 1, 2, and 3 (Milwaukee: Wisconsin Natural History Society and Milwaukee Public Museum, 1903), 12–13.

47. Undated note in TK papers. This story is repeated by several writers on Thure Kumlien, TK papers, box 1, folder 2.

48. Main, "Koshkonong Naturalist (III)," 323–24.

49. Brewer to Kumlien, October 25, 1873, TK papers, box 1, folder 1.

50. Brewer to Kumlien, October 25, 1873

51. Brewer to Kumlien, October 25, 1873.

52. Brewer to Kumlien, November 20, 1873, TK papers, box 1, folder 1.

53. Brewer to Kumlien, November 22, 1873, TK papers, box 1, folder 1.

54. Brewer to Kumlien, November 29, 1873, TK papers, box 1, folder 1.

55. Kumlien and Hollister, *Birds of Wisconsin*, 38.

56. Kumlien and Hollister, *Birds of Wisconsin*, 40.

57. A. W. Schorger, *Passenger Pigeon* 6, no. 1: 13–19, quoted in Sumner W. Matteson and R. Tod Highsmith, "Celebrating the Centennial of *The Birds of Wisconsin*: Remembering Ludwig Kumlien, with Selections from His 'Lost' Ledgers," *The Passenger Pigeon* 65, no. 4 (2003): 251.

58. Kumlien and Hollister, *Birds of Wisconsin*, 15–16.

59. Main, "Koshkonong Naturalist (III)," 340.

60. Main, "Koshkonong Naturalist (III)," 336.

61. Main, "Koshkonong Naturalist (III)," 336.

62. Adolphson, *A History of Albion Academy*, 68.

63. Main, "Koshkonong Naturalist (III)," 337.

64. Journal of T. V. Kumlien, donated by Kumlien descendant Anne Roseman to the Wisconsin Historical Society.

65. Kumlien to Brewer, December 23, 1874, TK papers, box 1, folder 1.

66. Journal of T. V. Kumlien.

67. Main, "Koshkonong Naturalist (III)," 335.

68. Brewer to Kumlien, April 23, 1875, TK papers, box 1, folder 1. We don't have Kumlien's letter of April 19.

69. Spencer Fullerton Baird, Thomas Mayo Brewer, and Robert Ridgway, *A History of North American Birds*, vol. 1 (Boston: Little, Brown, 1874), 565.

70. Baird, Brewer, and Ridgway, *A History of North American Birds*, 368.

71. Kumlien, Work Journal, June 25, 1875 (typescript by Albert O. Barton, Wisconsin Historical Society Archives, Madison).

72. Journal of T. V. Kumlien, September 22, 1875.

73. Hoard quoted in Angie Kumlien Main, "Studies in Ornithology at Lake Koshkonong and Vicinity by Thure Kumlien from 1843 to July, 1850," *Transactions of the Wisconsin Academy of Sciences, Arts and Letters* 37 (1945), 95. Hoard was later Governor of Wisconsin (1889–1891).

74. Journal of T. V. Kumlien, December 3, 1875.

75. Especially *The Wolfling*.

76. Undated fragment quoted in Main, "Koshkonong Naturalist (III)," 337.

Chapter 13

1. Captain C. J. Rollis, Stoughton, Wisconsin, to Angie Kumlien Main, May 10, 1920, scrapbook in TK papers, box 3.

2. Rollis to Main, May 10, 1920.

3. "Memoir of Thure Kumlien," September 6, 1888, Denver, Colorado, Alva Adams, governor of Colorado, scrapbook in TK papers, box 3. He was governor of Colorado at the time he wrote this memoir of Kumlien.

4. Rollis to Main, May 10, 1920.

5. Journal of T. V. Kumlien, donated by Kumlien descendant Anne Roseman to the Wisconsin Historical Society, May 18, 1875. "Dr. Hoy of Racine come here on a visit this evening. Ludwig brought him from the Depot."

6. "Mission, Vision & Values," About Us, Wisconsin Academy of Sciences, Arts and Letters website, www.wisconsinacademy.org/about-us/mission-vision-values.

7. Increase Lapham chaired the afternoon session at 2:30 on February 10, 1875, and read his paper on the law of embryonic development. There is no mention that Kumlien attended this session.

8. Latin names translated in text. *Transactions of the Academy of Sciences, Arts and Letters, vol.* III, 1875–1876, 56–57.

9. *Transactions* III, 1875–1876, 56–57.

10. Edward Asahel Birge to Thure Kumlien, September 17, 1879, quoted in Publius V. Lawson, "Thure Kumlien," *Wisconsin Academy of Sciences, Arts and Letters* 20 (1921): 682. Birge had studied with Agassiz at Harvard, was a limnologist who studied inland lakes and rivers, and was twice president of the University of Wisconsin.

11. Charles Mann to Kumlien, May 3, 1876, TK papers, box 2, folder 1.

12. Ludwig's widow is quoted in Mrs. H. J. Taylor, "Ludwig Kumlien" *Wilson Bulletin* (June 1937): 85.

13. Thure Kumlien to Theodore Kumlien, August 3, 1877, TK papers, box 1, folder 2.

14. Ludwig never did complete his degree at Madison but was later awarded a degree when he was a professor at Milton College.

15. Thure Kumlien to Theodore Kumlien, August 3, 1877. Because he was writing to Theodore, he included no scientific names.

16. Thure Kumlien to Theodore Kumlien, August 3, 1877.

17. Thure Kumlien to Theodore Kumlien, August 3, 1877.

18. Thure Kumlien to Theodore Kumlien, August 3, 1877.

19. Kumlien to Anderson, July 9, 1877, RBA papers, box 18.

20. William J. Park, *Madison, Dane County and Surrounding Towns: Being a History and Guide to Places of Scenic Beauty and Historical Note Found in the Towns of Dane County and Surroundings, including the Organization of the Towns, and Early Intercourse of the Settlers with the Indians, Their Camps, Trails, Mounds, etc.* (Madison, WI: W.J. Park & Co., 1877).

21. R. Tod Highsmith notes, "These are old hunter's names for various ducks: 'shelldrakes' likely refers to canvasbacks, although the term was also applied to the northern shoveler; the 'whistlers' are common goldeneyes; and 'butterballs' are likely the bufflehead."

22. Park, *Madison, Dane County and Surrounding Towns*, 629.

23. Park, *Madison, Dane County and Surrounding Towns*, 630.

24. Park, *Madison, Dane County and Surrounding Towns*, 630.

25. The Indian mounds, now called the Kumlien mounds, are still there—except for one removed in road building. Bjorkander didn't level them with the plow, nor did Thure or Theodore or the several generations of good stewards who have owned the land since.

26. Park, *Madison, Dane County and Surrounding Towns*, 630–31.

27. Park, *Madison, Dane County and Surrounding Towns*, 631.

28. Park, *Madison, Dane County and Surrounding Towns*, 631.

29. Henry Howgate, ed., *The Cruise of the Florence or, Extracts from the Journal of the Preliminary Arctic Expedition of 1877–78* (Washington, DC: James J. Chapman, 1879), 10.

30. Thure Kumlien to Theodore Kumlien, August 3, 1877.

31. Ludwig Kumlien and Ned Hollister, *The Birds of Wisconsin*, Bulletin of the Wisconsin Natural History Society, vol. 3, nos. 1, 2, and 3 (Milwaukee: Wisconsin Natural History Society and Milwaukee Public Museum, 1903), 39.

32. R. Tod Highsmith notes, "It is difficult to verify what bird Kumlien means here. The Northern pygmy owl (called pygmy owl in Kumlien's day) is a western bird and does not occur in Wisconsin (nor did it in Kumlien's day). Kumlien could have obtained one in a trade with another collector, but it's unlikely that he would have personally collected one from the Koshkonong area."

33. Ludwig Kumlien's ledger, quoted in Sumner W. Matteson and R. Tod Highsmith, "Celebrating the Centennial of *The Birds of Wisconsin*: Remembering Ludwig Kumlien, with Selections from His 'Lost' Ledgers," *The Passenger Pigeon* 65, no. 4 (2003): 282.

34. Kumlien and Hollister, *Birds of Wisconsin*, 11.

35. Kumlien and Hollister, *Birds of Wisconsin*, 33–34.

36. The Milwaukee Audubon Society was founded in 1897 in large part to stop the slaughter of birds at Horicon Marsh for their feathers and plumes.

37. Kumlien to Anderson, December 30, 1879, RBA, box 18.

Chapter 14

1. Kumlien to Spencer Fullerton Baird, September 6, 1880, photocopy from Susan Binzel family archive.

2. Edna L. Martin, "Memories of My Great Aunt Sophia Walberg," September 1969, typescript, family archive.

3. Undated notes by Angie Kumlien Main and letter to Ludwig Kumlien, TK papers, box 3, in scrapbook.

4. Quoted in Sumner W. Matteson and R. Tod Highsmith, "Celebrating the Centennial of *The Birds of Wisconsin*: Remembering Ludwig Kumlien, with Selections from His 'Lost' Ledgers," *The Passenger Pigeon* 65, no. 4 (2003): 256.

5. It's curious that Ludwig didn't pursue this job himself, since we know he was looking for full-time scientific work at that time. Or perhaps he did and was turned down.

6. Nancy Oestreich Lurie, *A Special Style: The Milwaukee Public Museum, 1882–1982* (Milwaukee: Milwaukee Public Museum, 1983), 14.

7. William Morton Wheeler, obituary, Sixth Annual Report of the Board of Trustees of the Public Museum of the City of Milwaukee, October 1, 1888.

8. Wheeler quoted in George Howard Parker, "Biographical Memoir of William Morton Wheeler," National Academy of Sciences Annual Meeting, 1938, p. 205.

9. A. W. Schorger, "The Wisconsin Natural History Association" *Wisconsin Magazine of History* 31, no. 2 (December 1947): 170.

10. Lurie, *A Special Style*, 6.

11. Martha Bergland and Paul G. Hayes, *Studying Wisconsin: A Life of Increase Lapham* (Madison: Wisconsin Historical Society Press, 2014), 311.

12. Record of Proceedings of the Board of Trustees of the Public Museum of the City of Milwaukee, April 21, 1884, p. 38.

13. Kumlien to Doerflinger, June 11, 1883, MPM Archives, Kumlien file.

14. "Fcw?" (perhaps F. C. Winkler) to Kumlien, December 15, 1883, MPM Archives, Kumlien file.

15. Kumlien to Doerflinger, December 18, 1883, MPM Archives, Kumlien file.

16. Telegram (copy) from Doerflinger to Kumlien, December 20, 1883, MPM Archives, Kumlien file.

17. Record of Proceedings of the Board of Trustees of the Public Museum of the City of Milwaukee, September 28, 1883, p.15.

18. Record of Proceedings of the Board of Trustees of the Public Museum of the City of Milwaukee, February 19, 1884, p. 38.

19. Wheeler quoted in Parker, "Biographical Memoir," 204.

20. Wheeler quoted in Parker, "Biographical Memoir," 204.

21. Lurie, *A Special Style*, 14.

22. Wheeler quoted in Parker, "Biographical Memoir," 203. R. Tod Highsmith describes Ward as "the superstore of natural history specimen dealers at the time."

23. MPM Second Annual Report, October 1, 1884, p. 16.

24. Doerflinger to Kumlien, June 25, 1884, Letter book, MPM Archives, Kumlien file.

25. Wheeler, quoted in Parker, "Biographical Memoir," 204.

26. Wheeler, quoted in Parker, "Biographical Memoir," 205.

27. Wheeler, quoted in Parker, "Biographical Memoir," 207.

28. Wheeler, quoted in Parker, "Biographical Memoir," 204.

29. Wheeler, obituary, 4.

30. Kumlien to Swea Kumlien Martin, undated, quoted in Angie Kumlien Main, "Thure Kumlien, Koshkonong Naturalist (III)," *Wisconsin Magazine of History* 27, no. 3 (March 1944): 341.

31. Wheeler, obituary, 5.

32. William Morton Wheeler, *The Flora of Milwaukee County* (Milwaukee: Wisconsin Natural History Society, 1884), 180, 181.

33. Wheeler, obituary, 5.

34. Wheeler, obituary, 5–6.

35. Kumlien to Martin, undated, quoted in Main, "Koshkonong Naturalist (III)," 340–41.

36. Kumlien to Doerflinger, April 30, 1885, MPM Archives, Kumlien file.

37. Kumlien to Doerflinger, April 30, 1885.

38. MPM, Third Annual Report, p. 9.

39. Thure Kumlien to Theodore Kumlien, October 4, 1885, in Main, "Koshkonong Naturalist (III)," 340.

40. Wheeler, obituary, 8.

41. MPM Third Annual President's Report, 9.

42. MPM Third Annual President's Report, 15.

43. Thure Kumlien to Theodore Kumlien, October 4, 1885, in Main, "Koshkonong Naturalist (III)," 340.

44. Main, "Koshkonong Naturalist (III)," 341.

45. Edward Lee Greene to Kumlien, December 22, 1885, quoted in Main, "Koshkonong Naturalist (III)," 334.

46. Edward Lee Greene, "Sketch of the Life of Thure Kumlien, A. M." in *Pittonia: A Series of Papers Relating to Botany and Botanists, 1887–1889*, vol. 1 (Berkeley, CA: n.p.), 259.

47. MPM Trustees Report, 1887, p. 7.

48. Main, "Koshkonong Naturalist (III)," 339.

49. Thure Kumlien to Frithiof Kumlien, no date, in Main, "Koshkonong Naturalist (III)," 339.

50. Thure Kumlien to Frithiof Kumlien, no date, in Main, "Koshkonong Naturalist (III)," 339.

51. Carl Doerflinger, October 1, 1887, Fifth Annual Report to the board of Trustees, 9.

52. Carl Doerflinger, Fifth Annual Report, 10.

53. Carl Doerflinger, Fifth Annual Report, 10.

54. Theodore Kumlien later had a stone made for the family, and on it he called his mother "Margretta."

55. Ludwig Kumlien to Carl Doerflinger, January 20, 1888, MPM Archives, Kumlien file.

56. Main, "Koshkonong Naturalist (III)," 341.

57. Main, "Koshkonong Naturalist (III)," 341.

58. Ludwig Kumlien to Edward Lee Greene, August 26, 1888, Greene Papers, University of Notre Dame Archives, copy from Kumlien family archives.

59. Ludwig Kumlien to Edward Lee Greene, August 26, 1888.

60. Ludwig Kumlien to Edward Lee Greene, August 26, 1888.

61. Wheeler, obituary, 6.

62. Main, "Koshkonong Naturalist (III)," 19.

Epilogue

1. Ludwig Kumlien to Edward Lee Greene, September 10, 1888, Greene Papers, University of Notre Dame Archives, copy from Kumlien family archives.

2. Angie Kumlien Main, *Bird Companions* (Boston: R.G. Badger, 1925). A later edition was published as *Handbook of Birds*.

Acknowledgments

During the four years I've worked on this book, so many people have helped. For this and for them I am very grateful.

Paul G. Hayes, Harry Anderson, Judy Punzel, and Susan Binzel contributed important materials and encouragement, which got me started on this project in 2016.

Thank you to the following people and institutions and associations in Sweden who contributed documents and details of Kumlien's life, friends, and specimens:

- Erland Bohlin, Biologiska museet, Karolinska gymnasiet, Örebro;

- Lars Bergdahl, ordförande Bjärka-Härlunda hembygdsförening, Axvall;

- Maria Asp, arkivarie, Centrum för vetenskapshistoria, Kungliga Vetenskapsakademien, Stockholm;

- Gunnar Eriksson, professor emeritus, Idé-och lärdomshistoria, Uppsala universitet, Uppsala;

- Mats Hjertson, museintendent, botanik, and Erica Mejlon, museintendent, zoologi, Evolutionsmuseet, Uppsala universitet, Uppsala;

- Jan-Gerhard Hemming, Härlunda;

- Inger Löfstedt, Håtuna-Håbo-Tibble forskargrupp, Bro;

- Rolf Jansson, Emmaboda;

- Kungliga biblioteket, Stockholm;

- Arne Anderberg, professor, botanik; Mattias Forshage, förste assistent, zoologi; Ulf Johansson, intendent, zoologi; Daniela C. Kalthoff, intendent, zoologi; Erik Åhlander, förste assistent, zoologi, Naturhistoriska Riksmuseet, Stockholm;

- Cilla Ingvarsson, samlingsregistrator, Sjöfartsmuseet Akvariet, Göteborg;

- Ingrid Ulfstedt, intendent, samlingsenheten, Sjöhistoriska Museet, Stockholm;

- Stifts-och landsbiblioteket, Forskaravdelningen, Skara;

- Stockholms Stadsbibliotek, kundtjänst, Stockholm;

- Marie Andersson, ordförande, Upplands fornminnesförening och hembygdsförbund, Uppsala;

- Camilla Lööf, antikvarie, samlingsavdelningen, Upplandsmuseet, Uppsala;

- Maria Berggren, chef för avdelningen för specialsamlingar, Åsa Henningsson, biträdande avdelningschef för avdelningen för specialsamlingar, Johan Sjöberg, arkivarie, avdelningen för handskrifter och musikalier, Uppsala universitetsbibliotek, Uppsala;

- and Torbjörn Walheim, Saltsjö-Boo.

Lena Peterson Engseth, researcher and translator, was essential to the Swedish chapters and translations of this book. Her knowledge of her native Sweden, its language and history and institutions, her poetic sensibility, and her persistence in an internet search, made her a joy to work with—a perfect researcher. It was Lena, nudging archivist Maria Asp at the Center for History of Science in Stockholm, who brought the story of Thure Kumlien's trip to Gotland Island to light after it had been missing for more than 170 years. I can never thank Lena enough for her research and for the pleasure she brought to this project.

I had help in many forms over the years of the project from Betty Adelman, Doug Armstrong, John Bergland & Carolyn Knox, Jim Bussey, Kathleen Dale, Gene & Jean Eisman, Susan Engberg, Ann Engelman, Susan Firer, Holly Hetzer, Anne Kingsbury & Karl Gartung, MaryLee Knowlton & Mark Sachner, Veronica Lundback, Barbara Minor, Kathy & Mike Mooney, Susan Moran, Randy Powers, Lisa & Mike Sattell, Carol Sklenicka, Jim Price, Kurt Sampson, Chuck Stebleton, Teri Sutton, Peter Thornquist, Lizzie Tuma, Wendy Walcott, and Susan Wooldrige.

Special thanks to Judy Schule of Milton College; Barbara Froemming, Bev Wenzel, and Jan Eherengren of the Swedish American Historical Society of Wisconsin; Merrilee Lee of the Hoard Historical Museum; Harvey Taylor and the Earth Poets; David Kottwitz of the Sterling North Society; Eric Baker at the Albion Historical Society; Ruth King at the Milwaukee Public Museum; the Milwaukee County Historical Society; and most especially Simone Munson and the archivists at the Wisconsin Historical Society Archives and the archivists at the Area Research Center Archives at University of Wisconsin–Milwaukee.

Thanks to Dorothy Uhrinak, Marianne Richie, Kathy Pearson, Aimee Jambor, and Sue Uhrinak for their generous support of color reproductions in this book.

R. Tod Highsmith elegantly clarified scientific bird names and caught many birdy errors. All remaining errors are mine.

Thank you to the many descendants of Thure Kumlien who have provided documents and information, especially to Susan Binzel, Gregg Kumlien, Anne Roseman, and Betsy D'Onofrio.

Very special thanks to Elizabeth Wyckoff of the Wisconsin Historical Society Press who patiently and professionally worked her considerable magic on the manuscript.

Thank you to my sister, Brita Bergland, who believed in this book on the days when I could not.

And thanks to my husband, Jim. The first time Jim Uhrinak and I drove down to Jefferson County and around Kumlien Country, Jim said to me, "Thure Kumlien kept his world larger than the acres he had under the plow." My naturalist husband then showed me that world as Kumlien himself might have—if he'd been driving a Nissan. We saw flatwoods, hidden springs, savanna and prairie remnants, swamp, mesic forest, southern flora, and more. Together, that first day, we found Kumlien's farm. With Jim, there is always discovery.

INDEX

Page numbers in *italics* refer to illustrations.

About the Author

Martha Bergland is the coauthor, with Paul G. Hayes, of *Studying Wisconsin*—a Society Press biography on famed Wisconsin naturalist Increase Lapham, which won the Milwaukee County Historical Society's Gambrinus Prize. She taught for many years at Milwaukee Area Technical College. She lives in Glendale, Wisconsin, with her husband, Jim Uhrinak.

PHOTO BY BARBARA J. MINER